TEACHING CARBON NEUTRAL DESIGN IN NORTH AMERICA

This book brings to light a diverse range of innovative architectural design studio methodologies formulated to educate future graduates to combat the climate crisis through carbon neutral design.

Award-winning professors detail tried-and-tested studio methodologies, outlining their philosophical rationale, the role of precedent study, design concept and professional partnerships, the approach to analytics and software design development, required readings, assignment and student work examples, and anticipated future innovation. Chapters are grouped under the varying focal points of community empowerment, bioclimatic response, performance analytics, design build, and urban scale, all adopting a holistic view of sustainable design that incorporates technical challenges as well as those of equity and social justice.

This heterogeneous compilation of strategies encourages wide accessibility to and acceptance by studio professors, as well as administrators and faculty developing architecture curricula. This will, in turn, maximize the impact on curtailing carbon emissions resulting from the construction and operations of our built environment.

Robin Z. Puttock, AIA, LEED AP BD+C, WELL AP is an assistant professor at Kennesaw State University and a practicing architect with 20 years of professional, national award-winning, sustainable design experience. Robin's research focuses on the pedagogy related to the built environment's role in both carbon neutrality and human well-being, with an emphasis on the connection between academia and the allied professions. She is the project architect of many LEED certified buildings and the first US Department of Education Green Ribbon School recognized by President Barack Obama and the US Department of Education. Robin serves as the 2025 chair of the National AIA Committee on the Environment (COTE) Leadership Group.

"The rapid decarbonization of American buildings is a formidable task facing the next generation of architects in North America. This thoughtful and comprehensive collection is an indispensable resource for educators seeking to equip architecture students with the knowledge and creativity they need to rise to this challenge, and it represents some of the innovative minds in our field today."

Lindsay Baker, *CEO, International Living Future Institute*

"Like meeting fire and life safety codes, the next generation of architects will be required to design and construct buildings free of greenhouse gas pollution. Getting there starts with the commitment of professors teaching architecture to change studio culture. This terrific compendium edited by Robin Puttock shows that teaching zero-carbon design is redefining design excellence."

Carl Elefante, *FAIA, FAPT, 2018 President American Institute of Architects, Senior Fellow Architecture 2030, Principal Emeritus, Quinn Evans*

"We must shift how we teach design to enable the radical changes in building design and construction needed to meet global climate goals. This impressive collection of innovative case studies is a critical resource to help develop, test, and refine curricula that will inspire and empower the next generation of architects and equip them to lead towards carbon neutral and ultimately climate-positive design solutions."

Kate Simonen, *Professor of Architecture, University of Washington, and Founder of Carbon Leadership Forum*

"Climate crisis is upon us, but academia has been woefully slow to adapt. This lively collection from Prof. Puttock goes a long way towards correcting that, providing both teachers and students a framework for design in a difficult new time. Highly recommended for everyone in architecture."

Bruce King, *PE, Director, Ecological Building Network*

"*Teaching Carbon Neutral Design in North America* is a transformative resource featuring contributions from leading educators in the field. Within 20 chapters, it covers essential aspects of sustainable design from exploration of passive design in diverse climates, to championing shared energy paths in communities, to insights on equity and community empowerment – all innovative approaches revealing the importance of partnering with real-world clients and providing students with real hands-on experience in addressing the various sustainability measures. This is a must-read for educators and practitioners dedicated to advancing sustainable architecture and addressing climate change."

Alison Kwok, *Professor of Architecture, University of Oregon and Director of the NetZED Laboratory*

"Robin Puttock has designed a timely volume of articles that shares innovative didactics for a carbon neutral architectural future. The contributions in this book cast environmentally responsive teaching beyond embodied and operational considerations toward architecture's complex socio-technical entanglements of carbon, energy, resource circularity, practice, equity, justice, health, and community. This is a valuable reference on teaching pedagogy for 21st-century architectural educators."

Ryan E. Smith, *PhD, Director and Professor, University of Arizona, Founding Partner of MOD X*

"The climate emergency underscores the importance of teaching sustainable, low-carbon design to architecture students. This book demonstrates how this can be done with a well-selected compilation of teaching experiences taught by award-winning practitioners and educators. It is an excellent reference book that can be used by educators worldwide."

Pablo La Roche, *PhD, Professor of Architecture, Cal Poly Pomona University and Principal, Arcadis*

TEACHING CARBON NEUTRAL DESIGN IN NORTH AMERICA

Twenty Award-Winning Architectural Design Studio Methodologies

Edited by Robin Z. Puttock

LONDON AND NEW YORK

Designed cover image: Robert Conway and Morgan Weber

First published 2025
by Routledge
4 Park Square, Milton Park, Abingdon, Oxon OX14 4RN

and by Routledge
605 Third Avenue, New York, NY 10158

Routledge is an imprint of the Taylor & Francis Group, an informa business

British Library Cataloguing-in-Publication Data
A catalogue record for this book is available from the British Library

ISBN: 978-1-032-69254-8 (hbk)
ISBN: 978-1-032-69251-7 (pbk)
ISBN: 978-1-032-69256-2 (ebk)

DOI: 10.4324/9781032692562

Typeset in Sabon
by Taylor & Francis Books

**Dedicated to my husband, Mark Puttock,
for his unwavering support that made my transition
into academia (and this book) possible**

CONTENTS

FIGURES

TABLES

CONTRIBUTORS

Omar Al-Hassawi, PhD is an associate professor at Washington State University's School of Design and Construction. Omar's teaching has been internationally recognized by the Association of Collegiate Schools of Architecture and the American Institute of Architecture Students who awarded him with the 2021 New Faculty Teaching Award.

Kjell Anderson, FAIA, CSBA, LEED Fellow is LMN's director of sustainable design and leader of LMN's Green Team, spearheading initiatives into energy modeling, materials + health, and water use reductions. Kjell is an international sustainability expert with 20 years of experience and author of *Design Energy Simulation for Architects*, published by Routledge in 2014.

Dorit Aviv, PhD, AIA is an assistant professor of architecture at the Weitzman School of Design at the University of Pennsylvania, specializing in sustainability and environmental performance. She is the director of the Thermal Architecture Lab, a cross-disciplinary laboratory at the intersection of thermodynamics, architecture and material science.

Jonathan Bean, MS, PhD, CPHC is associate professor of architecture, sustainable built environments, architectural engineering, and marketing at the University of Arizona, a joint appointee at the National Renewable Energy Laboratory, and the co-director of the University's Institute for Energy Solutions. He has served as faculty lead to 13 Solar Decathlon Design Challenge finalist teams, including the 2022 Commercial Grand Prize winner.

William W. Braham, PhD, FAIA is a professor of architecture at the Weitzman School of Design at the University of Pennsylvania where he is director of the

Environmental Building Design programs and of the Center for Environmental Building + Design. He has worked on energy and architecture for over 40 years as a designer, consultant and researcher.

Craig Brandt, FAIA, AFAAR, LEED BD+C is an architect based in Chicago and adjunct professor at the University of Notre Dame. His teaching focuses on adaptive and regenerative design projects as well as cross-disciplinary design processes in arts and design. His practice in Chicago centers around civic, government, and higher education projects in various contexts.

Nancy Yen-wen Cheng, RA, LEED AP, NOMA teaches architecture and digital methods at University of Oregon (UO). She researches design methods, currently timber design for disassembly and reuse for adaptive housing. Cheng has led ACADIA, the AIA Technology in Architectural Practice group, and the UO Architecture Department and Portland Program.

Erica Cochran Hameen, PhD, associate AIA, NOMA, LEED AP is the director of DEI, co-director of the Center for the Building Performance and Diagnostics, track chair of the Doctor of Design (DDes) program, and an associate professor at Carnegie Mellon University. Erica received numerous honors for her research, teaching, and designs focused on sustainability, indoor environmental quality, energy efficiency and policy.

Tom Collins, PhD, AIA, LEED AP is an associate professor of architecture at Ball State University. He teaches high-performance building design, environmental systems, and sustainability. He has advised 25 Solar Decathlon Design Challenge teams and was faculty co-lead for Ball State's 2023 Solar Decathlon Build Challenge entry, which won the first place prize overall.

Edoarda Corradi Dell'Acqua, MS is an associate teaching professor in the Department of Civil, Architectural, and Environmental Engineering at Illinois Institute of Technology, where she also serves the Ed Kaplan Family Institute for Innovation and Tech Entrepreneurship. She was recognized with Illinois Tech's John W. Rowe Award in 2021 and the Solar Decathlon Richard King Award in 2023.

Claude M. H. Demers, MArch, PhD graduated from the University of Cambridge. Her research innovates on the visualization of qualitative and quantitative assessments of daylighting in architecture. Co-founder of the Groupe de Recherche en Ambiances Physiques at Laval University, she develops biophilic design methods associating the importance of wellbeing through bioclimatic design.

Timur Dogan, PhD is an associate professor and MArch program director at Cornell AAP. Dogan holds a PhD from MIT, an MDES from Harvard GSD, and a Dipl-Ing in architecture with distinction from the TU Darmstadt. Dogan's research aims to accelerate building decarbonization through educational programming and strategic research at the intersection of design, computer science, building performance simulation, and urban geospatial analysis.

Kelly Alvarez Doran, OAA, MRAIC, Architecture 2030 senior fellow is a father, architect, educator, and activist. His holistic approach to the design of the built environment has been shaped by his experiences working across the world. Kelly, a co-founder of Ha/f Climate Design, is a regular speaker, writer, and advocate for the integration of life cycle assessments into design thinking.

Billie Faircloth, FAIA, LEED AP BD+C is an architect, educator, and transdisciplinarian who has transformed practice-integrated research and earned a national and international reputation for demonstrating its value, methods, and outcomes. Billie is an associate professor at Cornell University and a senior faculty fellow at the Cornell Atkinson Center for Sustainability.

Lee A. Fithian, PhD, AIA, NCARB, LEED AP is a practitioner, educator, researcher, and director of the OU Online MS in sustainable architecture which provides technical skills in the digital world of climate change, sustainability, and building performance analysis. She holds a patent for novel building facade technology that enhances air quality in urban street canyons.

Rob Fleming, AIA, NOMA, LEED AP is an award-winning educator, author, LEED accredited professional, and a registered architect. He is the director of the Center for Professional Learning at the Weitzman School of Design at the University of Pennsylvania, where he develops accessible and affordable educational programs focused on sustainable design and design leadership.

David Franco, PhD, RA is an associate professor and director of the School of Architecture at the University of Texas at Arlington. He holds a PhD in architectural history and theory and an MArch from the Universidad Politécnica de Madrid. His research focuses on political and societal issues in modern architecture.

Robert Fryer, RA, LEED AP, NCI is the director, professor, and cofounder of the MS Sustainable Design Program at Thomas Jefferson University. He is an award-winning educator, architect, and author. He is writing a book on resilient and adaptive design pedagogy. Robert graduated from the Architectural Association and is a registered architect.

Ulrike Heine, Dipl-Ing is an associate professor of architecture at Clemson University and teaches design studio and technical seminars in the Graduate Program. She graduated with Diplom Ingenieur from the Brandenburg

Technical University in Cottbus (Germany). Her research and scholarship focuses on affordable, environmentally responsive design in a local context.

Julie Ju-Youn Kim, RA, AIA, NCARB is the William H. Harrison Professor and Chair at the Georgia Tech School of Architecture, where she directs the Flourishing Communities Collaborative. As an educator and a licensed architect, Julie leverages data-driven and quantitative methodologies to solving social and cultural problems in the built environment.

Nea Maloo, FAIA, NOMA, LEED AP, ICC is an assistant professor at Howard University where she teaches environmentally sustainable systems, decarbonization, and professional practice. Maloo is also the founder and principal architect of Showcase Architects, a firm distinguished for its innovative approach to environmental architecture, carbon planning, and biophilic design.

Elizabeth Martin-Malikian, associate AIA, NOMA is a nonprofit leader and educator. From 2021–2023, she served as the chief executive officer (CEO) and executive director of the Cosanti Foundation, an Arizona-based 501(c)(3) nonprofit organization with a mission to inspire a reimagined urbanism that builds resilient and equitable communities sustainably integrated with the natural world.

Ulrike Passe, Dipl-Ing, Architekt, AIA International Associate is a professor of architecture at Iowa State University and focuses on integrative sustainable design, natural ventilation, and microclimate. She is a co-author of *Designing Spaces for Natural Ventilation*, and her students have won the ACSA/AIA COTE Top Ten competition six times. Her research is funded by the US National Science Foundation.

Nihar Pathak, LEED AP BD+C is an AECM PhD candidate, graduate instructor, and UDream teaching fellow at Carnegie Mellon University. Her research delves into life cycle impact assessments for construction materials, reshaping the AECM industry. She also focuses on enhancing indoor environmental quality, redefining architecture and STEM education.

André Potvin, MArch, PhD graduated from the University of Cambridge. He is a professor, actively involved in research and teaching environmental design at graduate and postgraduate levels. Dr Potvin cofounded the Groupe de recherche en ambiances physiques (GRAP) dedicated in passive environmental control strategies urban microclimatology and environmental adaptability from intermediate spaces.

Robin Z. Puttock, AIA, LEED AP BD+C, WELL AP is an assistant professor at Kennesaw State University and a practicing architect with 20 years of professional, national award-winning, sustainable design experience. Robin's research focuses on the pedagogy related to the built environment's role in both carbon neutrality and human well-being, with an emphasis on the connection between academia and the allied professions. She is the project architect of many LEED certified buildings and the first US Department of Education Green Ribbon School recognized by President Barack Obama and the US Department of Education. Robin serves as the 2025 chair of the National AIA Committee on the Environment (COTE) Leadership Group.

George Schafer, PhD, RA is an architect with BOUDREAUX in Charlotte, North Carolina, and was a senior lecturer of architecture at Clemson University between 2015–2023. He holds a PhD in planning, design and the built environment from Clemson and an MArch from Harvard. His research focuses on human-centered design and interactive environments.

Joseph Wheeler, AIA is a professor of architecture at Virginia Tech and co-director of the university's Center for Design Research. He is a globally recognized innovator in the areas of housing industrialization, environmental sustainability, and technology integration. Wheeler has spent two decades helping student researchers and industry, government, and academic partners explore solutions to the world's global housing challenges.

Yang Yang, PhD is an assistant professor at Thomas Jefferson University. She holds a PhD and MSAUD from Cornell, an MArch and BEng from Tongji University. Yang is a designer and engineer specializing in urban data analytics and simulation modeling. She is experienced in computational design and the development of design-aiding software tools.

PREFACE

In 2015, with the financial and emotional support of my husband, I decided to take a professional leap. I sold my partnership shares in my architecture practice and went back to graduate school with the goal of becoming an architecture professor. My bachelor of architecture from Virginia Tech had served me quite well in my 20-year professional career as a civic architect specializing in sustainable design. However, if I wanted to teach full-time, I would need a post-professional master's degree. I was accepted into Virginia Tech's Master of Architecture program and, as luck would have it, sixteen months later, I received my diploma in the mail, along with my first paycheck as an adjunct faculty member. Three-and-a-half years after that day, I had become a full-time assistant professor of practice and an associate dean, and was creating courses that I was passionate about, one of which was the Vertical Net-Zero Design Studio which is the inspiration for this book.

When I entered academia from practice, I was familiar with the design strategies and analytics required to achieve decarbonization. However, I was not quite sure where to begin when creating an architecture studio curriculum. Fortunately, as an AIA member, I was familiar with the AIA Framework of Design Excellence and, more specifically, the curriculum that the National AIA COTE (Committee on the Environment) Leadership Group (LG) created for the AIA COTE Top Ten for Students Competition. This curriculum aligned with my philosophy of carbon neutral design excellence and I adopted it into my integrated studio and net-zero studio pedagogies.

After my students won the competition in 2021, I joined the National AIA COTE Working Group for the student competition to see how I could support outreach efforts. The next year, I was appointed to the LG and continued to advocate for academic issues related to the environment. As a result, my academic network grew and I became aware that decarbonization pedagogies were being used in unique and exceptional ways by dedicated professors across North America. However, I also learned that there were many schools and professors who were not yet teaching this content. It made me think that these unique and exceptional strategies could be promoted and celebrated in a collaborative effort with the goal of increasing adoption of decarbonization pedagogies in architecture schools worldwide. Thus, the idea for this book was born!

INTRODUCTION

Robin Z. Puttock

There are hundreds of architecture programs across higher education institutions in North America; however, not all of them place an emphasis on pedagogy related to educating future graduates to combat the climate crisis through carbon neutral design. In an effort to increase the adoption of this content into curricula across the continent, this book was conceived to highlight the many existing diverse and innovative methodologies, each detailed in a chapter written by an international award-winning professor in this field.

The chapters have been grouped thematically into five parts: "Community Empowerment," "Bioclimatic Response," "Performance Analytics," "Design Build" and "Urban Scale." However, it must be noted that few, if any, of the chapters can be described solely by this categorization. The pedagogies illustrated in this volume are rich and, as a result, they often weave together several of the themes listed above. The final chapter placements are the result of conversations between the editor and each of the contributors, reflecting the mutual understanding of the foremost focus of the course.

All chapters follow a framework addressing five main topics:

1. course overview;
2. course methodology;
3. outreach opportunities;
4. course assessment; and
5. looking ahead.

The goal of this heterogeneous compilation of strategies is to encourage wide accessibility to and acceptance by faculty and administrators developing architecture curricula. It is the hope that this will, in turn, maximize the impact on

DOI: 10.4324/9781032692562-1

curtailing carbon emissions resulting from the construction and operations of our built environment.

Course Overview

The first section of each chapter includes general information describing each course and the intended learning objectives and learning outcomes. The intent for this section is for the reader to quickly determine if this chapter will be applicable to their needs, i.e. the level of students, the degree program, the goals of the course in terms of student learning, as well as the instructor information and university location.

Course Methodology

The next portion of each chapter describes the author's course methodology. This includes each instructor's unique philosophy as it relates to how we, as educators, see our impact on the future of carbon emissions resulting from the design, construction and operations of the built environment. In order to facilitate a connection between the reader and the most applicable content, it is important that the reader quickly understands the perspective of each contributor and can make a determination if the rest of each chapter will be a good fit in their studios.

An effective way to streamline this determination is through the unique definition of "carbon neutrality," provided by each chapter contributor. Similar to the word "sustainability," "carbon neutrality" can mean vastly different things to different people and it is critical we start from a clear contextual definition before diving into each course methodology. In addition, each author has defined other terms pertinent to their chapters in this section. This will alleviate the need for flipping back and forth to a volume-wide glossary that may not be as helpful or chapter-specific.

After the definition section, each contributor describes their course implementation. In discussions with the contributors prior to greenlighting the writing of the chapters, it was agreed that this portion of the book should be granted some flexibility and freedom from the otherwise rigid structure. Therefore, you will see a variety of sub-topics in this section including: the role of precedent and site analysis, professional partnerships, assignments, analytics and required and recommended readings. Student work examples are typically interspersed throughout the chapters as required to support the chapter content. All of the above implementation strategies are provided as a buffet for readers to pick and choose from, based on those that align with their personal philosophical approach to teaching.

It is important to note that this book, whose main goal is the sharing of pedagogical approaches, may not have been possible ten years ago. In my experience, there has been a shift due to the intensity of the climate crisis and we are now in an era of general agreement that sharing information is of critical importance when attempting to have an impact on this wicked problem. I encourage readers to contact any of the authors featured in this volume with follow-up questions or discussion points. Each contributor, by agreeing to participate in this publication, has proclaimed their belief that we are in this together and we must work together.

Outreach Opportunities

This section on outreach is the result of contributor discussions prior to the commencement of writing. It was included because of our fundamental belief that this work cannot be done in silos. We must share the ideas and the innovations. There are infinite ways to do this, and each chapter suggests unique and creative ways of doing just that. It is our hope that readers implement these ideas, and build upon them. The sky is the limit and we must share now.

Course Assessment

As educators, we acknowledge the value of assessment. However, the term "assessment," similar to the term "carbon neutral," can mean vastly different things to different people. Therefore, this section is included in an effort to celebrate and proliferate the many different approaches to assessment, with the agreed upon understanding that some form of assessment is critical when designing a course. Authors describe a variety of assessment criteria including the National Architectural Accrediting Board (NAAB) 2020 Conditions for Accreditation, course rubrics, competition outcomes, student course evaluations, measurement of student learning outcomes and general lessons learned.

Looking Ahead

This volume presents the views of 26 authors within 20 chapters. We agreed to close each chapter with a critical reflection on where we are, where we are headed and where we need to be. Given the nature of the climate crisis, it is understood that some of these reflections will be cautionary. However, given the mission of the assembled authors who have shared their time and talent contributing to this book, tirelessly working to disseminate information for the collective good, these reflections are also meant to be accessible and optimistic calls to action.

We hope this collection of diverse architectural studio methodologies inspires, encourages and energizes you for the semesters to come.

PART I
Community Empowerment

1

THE NET-ZERO VERTICAL DESIGN STUDIO

A Framework for Design Excellence

Robin Z. Puttock

Course Overview

Course Name

ARCH401, ARCH601, ARCH701 Net-Zero Vertical Design Studio

University / Location

The Catholic University of America / Washington, DC, USA

Targeted Student Level

- Undergraduate
  - 4th year, Bachelor of Science in Architecture, 4-year program
- Graduate
  - 1st year, Master of Architecture, 2-year program
  - 2nd year, Master of Architecture, 3-year program

Course Learning Objectives

In this course, students will:

- Analyze and synthesize data into design information.
- Pursue, through a rigorous quest, the evolution of their initial ideas.

DOI: 10.4324/9781032692562-3

- Bring together their knowledge and skills in offering solutions as they relate to the challenges of the built environment.
- Communicate their ideas and concepts verbally, in writing and visually.

Course Learning Outcomes

Successful completion of this course will enable students to:

- Integrate the concepts of various building systems and sub-systems into a design solution.
- Work using a mature method that encompasses all aspects of architectural design including conceptualization, research, development, technical integration, and presentation.
- Use a variety of building performance analytic software programs to design high performance architecture.
- Communicate with a variety of allied professionals in conversations related to their designs.

Course Methodology

Philosophy

The goal of this course is to inspire students to incorporate decarbonization strategies to the fullest extent possible in their work and future careers. In order to do this, the instructor must meet each student at their current level of understanding at the beginning of the course and enthusiastically empower the students to engage in the rapid innovation of the movement. There is also a psychological component to this course. This is typically one of the most challenging studios that the students take in their academic career, and in order to successfully receive the course content, and eventually apply it to their professional work, the students need to build on their understanding of how to work effectively in teams. Thus, the course not only requires collaboration but demonstrates how it is a part of effective practice to achieve decarbonization.

Definition

Carbon neutrality – No net release of carbon dioxide into the atmosphere as the result of the construction or operation of the built environment.

Implementation

General

The studio is a vertical option studio, meaning it is a studio comprising both undergraduate and graduate students and is one of several courses offered to satisfy the ARCH 401, 601 or 701 requirement. As a result, the course typically attracts students who are highly motivated to learn about carbon neutral design. The course content is delivered in a variety of ways throughout the semester including selected readings and videos, software tutorials and quizzes, mini-lectures (called 101s) at the beginning of every class for the first 8 weeks, guest speakers, precedent analysis, site visits, tours and charrettes, active design workshops and desk crit and guest juror pinup feedback. Students work as individuals and in groups. Every studio day is highly choreographed and the requirements are communicated clearly to the students well in advance as there is much to cover and the students come to the course with a wide range of expertise in regards to carbon neutral design.

Assignments

The 15-week semester is divided into 6 parts, each with a series of deliverables:

- Week 1: Philosophy
- Weeks 2–3: Pre-Design Analysis
- Weeks 4–6: Concept Design
- Weeks 7–9: Schematic Design
- Weeks 10–11: Design Development
- Weeks 12–15: Representation

Before diving into the 15-week curriculum, it is important to note that students are asked to familiarize themselves with the selected course software prior to the start of the semester. The required software varies each time this course is taught; however, it typically includes programs that deliver EUI calculations, daylighting modeling and embodied carbon and life cycle analysis. Students are provided with carefully curated one pagers of web links to specific tutorials for each software to support this endeavor. Quizzes are given early in the semester to ensure students are familiar with the software basics before the design work begins.

Transitioning into the actual course design, the first week of the semester is called "Philosophy" and is designed to introduce the students to why this course is critical for future designers of the built environment and to understand their role in addressing the climate crisis. Students are asked, as individuals, to analyze examples of design excellence through precedent study of previous American Institute of Architects (AIA) Committee on the Environment (COTE) Top Ten Professional winning projects and AIA COTE Top Ten for Students award-winning projects. In addition, students are required to carefully examine a matrix of ten design principles and associated design questions to consider for each principle (Table 1.1). This matrix is adapted from the AIA COTE Top Ten for Students competition website, which is, in turn, adapted from the AIA's Framework for Design Excellence. The ten principles in the matrix are: Integration, Equitable Communities, Ecosystems, Water, Economy, Energy, Well-being,

TABLE 1.1 A portion of the Framework for Design Excellence Matrix, only Principle 7: Well-being shown.

Principle	*Description*	*Narrative*	*Deepen and Refine*
Design for **WELL-BEING**	Sustainable design creates **comfort, health,** and **wellness** for people who inhabit or visit buildings	Discuss design strategies for optimizing daylight, indoor air quality, connections to the outdoors, and thermal, visual, and acoustical comfort	How does design promote the **health** of the occupants? Design **strategies** for daylighting, task lighting, views, ventilation, indoor air quality, and personal control systems
May Include	*Suggested Graphics*	*Metric*	*Studio Opportunities*
How does design promote the **health** of the occupants? How does design promote **activity** or exercise, access to healthy food choices, etc. Outline of **material** health strategies, including selection strategies Design strategies for day**lighting**, task lighting, and views Design strategies for ventilation, indoor **air** quality, and personal control systems How the project's design enhances users' connectedness to **nature** Design team approach to **integration** of natural systems and appropriate technology	**Model** photos, drawings or diagrams of daylight and ventilation strategies; test models	**Percent** of the building that can be daylit (only) during occupied hours; Percent of floor area with views to the outdoors; Percent of floor area within 15 ft. of an operable window. Daylight performance using the following concepts: Daylight Availability, or Annual Sunlight Exposure along with Spatial Daylight Autonomy: % of regularly occupied area achieving at least 300 lux at least 50% of the annual occupied hours	Daylight Calcs WELL Diagram Biophilia Graphic

Source: The author, adapted from the AIA Framework for Design Excellence

Resources, Change and Discovery. Students are then tasked, as individuals, with selecting 3–4 principles that they would like to focus on during the 15 weeks. This task not only requires a careful first review and familiarization of "The Framework", but it also provides some insight into aligned interests of their peers and potential partner selection at the end of week 3 of the semester.

Also, during this first week, comes significant content delivery. One of the main goals is to ensure all students in the studio are familiar with the basic concepts related to decarbonization strategies. Because this is a vertical studio, students are quite varied in their understanding of decarbonization and the strategies that a design can employ to achieve it. Therefore, in an effort to ensure all students have at least a foundational understanding, several mini-lectures, or 101s, are delivered this first week including Carbon 101, Materials/Reuse 101, LCA 101, Precedent/Site/Vicinity Analysis 101 and Existing Conditions/Program Analysis 101. These mini-lectures are 10 minutes or less each and often include short videos (*see Videos heading*). This has proven to be a successful method of content delivery, per the course evaluation feedback. Students appreciate the video format and hearing from additional voices beyond their instructor. This first week is of critical importance. The students need to feel that there is an ethical imperative to this work. The more passion displayed by the instructor during this first week, the better the student learning outcomes will be by the end of the semester.

The second and third weeks of the semester are introduced as "Pre-Design Analysis" and typically include a visit to the site(s). Each semester, the studio partners with a real-world client on a real project. Recent examples include a feasibility study of five adaptive reuse sites for Arlington Public Schools, a rapidly growing school system in Northern Virginia, and design options for a triangular city block of existing buildings in the Shaw neighborhood of Washington, DC. Depending on the project brief, guided tours of the site are provided by a variety of project stakeholders including city/county planners, adaptive reuse experts, owners' representatives and community members. This is an effective way for the students to begin to engage the members of the community that they will be serving. These tour guides also typically serve as jurors throughout the semester.

During these two weeks, students focus on four main topics:

1. applying what they learned during the first week of precedent study to their site analysis work, which is done in groups according to student interest in each of the site analysis tasks;
2. further developing their understanding of the different software, with the assistance of a graduate assistant and assessed with quizzes that they must repeat until they earn at least 80%;
3. learning more about various climate design and resilience strategies for their site via in-class 101s, and perhaps most importantly for this community-outreach studio; and
4. collecting community input data via in-person and online interviews, phone calls, emails and online surveys.

FIGURE 1.1 Student site visit in the Shaw neighborhood of Washington, DC, with tour guide and adaptive reuse expert, Carl Elefante.
Source: Photograph by the author

Depending on the studio's community partner, for example, Arlington Public Schools, the studio may begin with a highly prescriptive program. The community input is used in a similar way to how it is used in practice, to make suggestions to the owner in regard to the existing given program. Likewise, in another studio with a different community partner, for example, DC Office of Planning, the client might be completely open to student ideas in regard to the programs for the adaptively reused city block or a portion of the city block. This community research portion of the studio is a highly significant aspect of the pedagogy as it allows the students to build their soft skills related to community outreach, to understand different methods of data collection and to witness how a design can be impacted by community feedback. The students are asked to reflect upon the community feedback and compare it to the ten principles in the matrix that they studied early in the semester. What principles are most important to the community? How might this inform the design solutions moving forward?

At the end of this Pre-Design Analysis phase, and now having completed three weeks of studio coursework, students have a good sense of who their peers are, their interests and their work ethics. From this point forward, the students will be working in teams. To aid the selection process, the students complete a team-forming questionnaire, which has been meticulously honed over the years in an effort to maximize skill representation on each team and minimize personality conflicts among team members for the remaining 11 weeks of the course. Teams are typically 2–3 students, 2 being preferable.

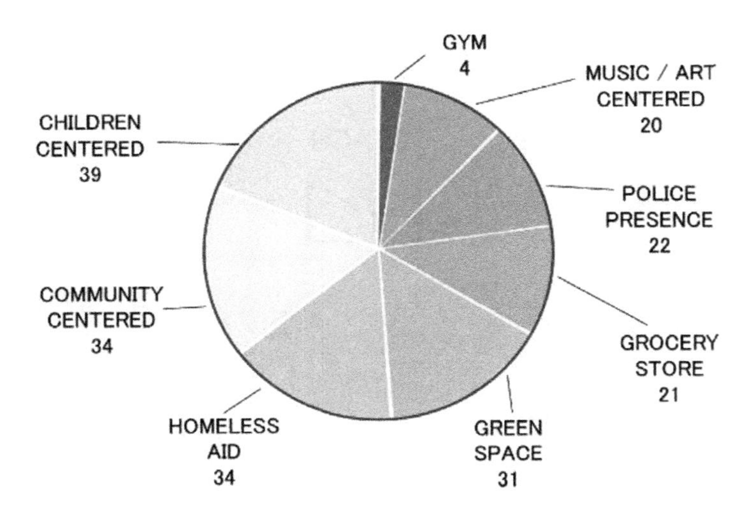

FIGURE 1.2 Community outreach data map and analytical summary.
Source: Student work by Katherine Kaderabek and Melina Skinner

Once the teams have been formed, week 4 begins the three-week "Concept Design" phase. This phase kicks-off with a team-building charrette informed by a Concept 101. Each team brainstorms a variety of concept words, phrases, parti drawings and models. The more the better. The format is open and loose. The day is meant to be fun. At the end of the day, each team selects three concepts that they would like to move forward. As this phase progresses, each day the teams are required to add parameters to their concepts including basic building design strategies from *Heating, Cooling, Lighting* (see Required Reading below), passive building and site strategies related to the sun and wind, potential reuse scope and strategy, building and site program needs, biophilia, WELL features, water harvesting and stormwater management. Each of these parameters are introduced with a 101. Because most of this content reinforces the relationship between building and site, the studio hosts a site charrette mid-way through this phase with landscape architects and civil engineers who offer feedback on preliminary site designs.

The concept design phase culminates with a juried presentation during which discussions focus on the selection of a single concept to move forward into schematic design. Special attention is given to how each concept addresses the ten principles in the Framework for Design Excellence. After the presentations, each student is required to submit a self-evaluation form. This form has been honed over the years, much like the team selection form, in an effort to identify any interpersonal or intrapersonal issues that may be developing which may affect a student's positive studio experience.

Schematic Design is the next three-week section of the studio. This phase continues to focus on passive design strategies, while also incorporating active design strategies, life safety codes, accessibility requirements, cost estimation strategies and envelope design options, each of which have at least one associated 101. This portion of the studio kicks off with an active design workshop during which

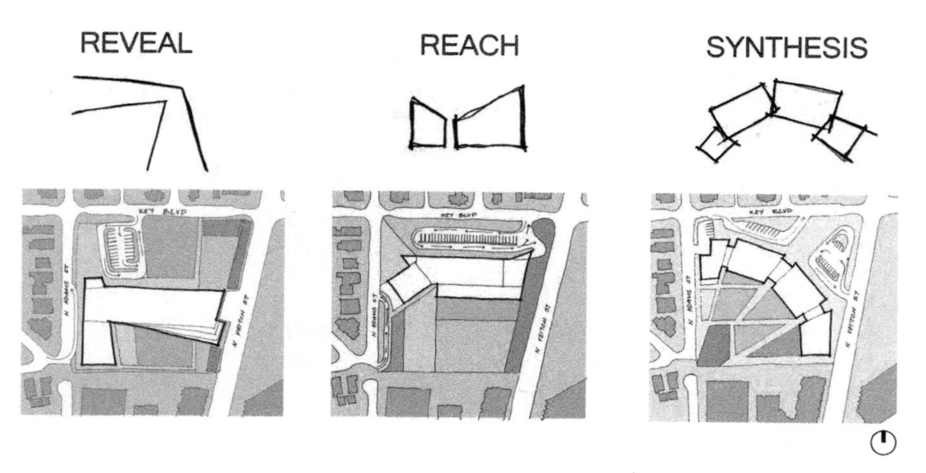

FIGURE 1.3 Three concept options at the conclusion of Concept Design.
Source: Student work by Mara Walsh, Maria Short and Emily Navarro

invited mechanical engineers and commissioning agents rotate between each group to learn about the group's designs and then make suggestions related to the pros and cons of each potential active design option. It is important that the students understand the valuable role these professionals play within the design team. Each semester in the course evaluations, the students consistently report this event to be one of their favorite activities in the studio and admit their amazement at learning how creative mechanical engineers and commissioning agents can be. During this phase, in addition to 2–3 desk crit visits each with outside guest reviewers (owner's representatives and subject matter experts like Bill Browning) the students continue to advance their schematic design. Also, during this phase, each student is required to reflect upon how their work has addressed each of the ten principles of The Framework for Design Excellence by once again reviewing the matrix from the first week of the semester and writing 100-word first drafts about each principle, as well as an overall narrative describing the main themes of their project. Schematic Design culminates in another juried review. This review is typically done on Zoom with a slide presentation for two reasons. First, this allows the jury to be from diverse geographical locations, specifically selected for their expertise as related to each student project and, second, this format allows students to practice their soft skills of technical preparation and online presentation. Once again, after the presentations, each student is required to submit a self-evaluation form.

One of the main goals of this studio is to attempt to mimic professional practice in as many ways as possible. Themes of teamwork, interdisciplinary collaboration, integrated design and iteration are consistently incorporated throughout the semester as well as illustrations of project management, scheduling and critical path. These themes all culminate in the next phase: Design Development. Students are surprised when they hear that the big design decisions are to be finalized by the end of the Schematic Design phase, week 9,

FIGURE 1.4 One of the student groups during the Active Design Workshop rotation with guest Roger Frechette (left) and student desk crits with DC Office of Planning's Timothy Maher (right).

Source: Photographs by the author

which is only 2/3 of the way through the semester (*this is generally in stark contrast to the studios they have had earlier in their academic careers*). After the schematic design review, the design is simply to be developed.

One of the most effective strategies to keep the students focused on minor revisions in lieu of major re-designs is to encourage a continued focus on analytics. What minor envelope selections can be made to lower the EUI? What tweaks can be made to the fenestration to improve the daylighting design? The human experience? Are there ways to increase the number and/or efficiency of photovoltaic panels? Can the stormwater management design be improved? Targeted investigations with clear goals are effective strategies to suggest at this point in the semester. This phase of the studio once again culminates in a presentation, typically in person, and one more self-evaluation form for the final push.

The final 3–4 weeks, depending on the assigned final presentation date, are dedicated to the curated representation of the work completed during the first 11

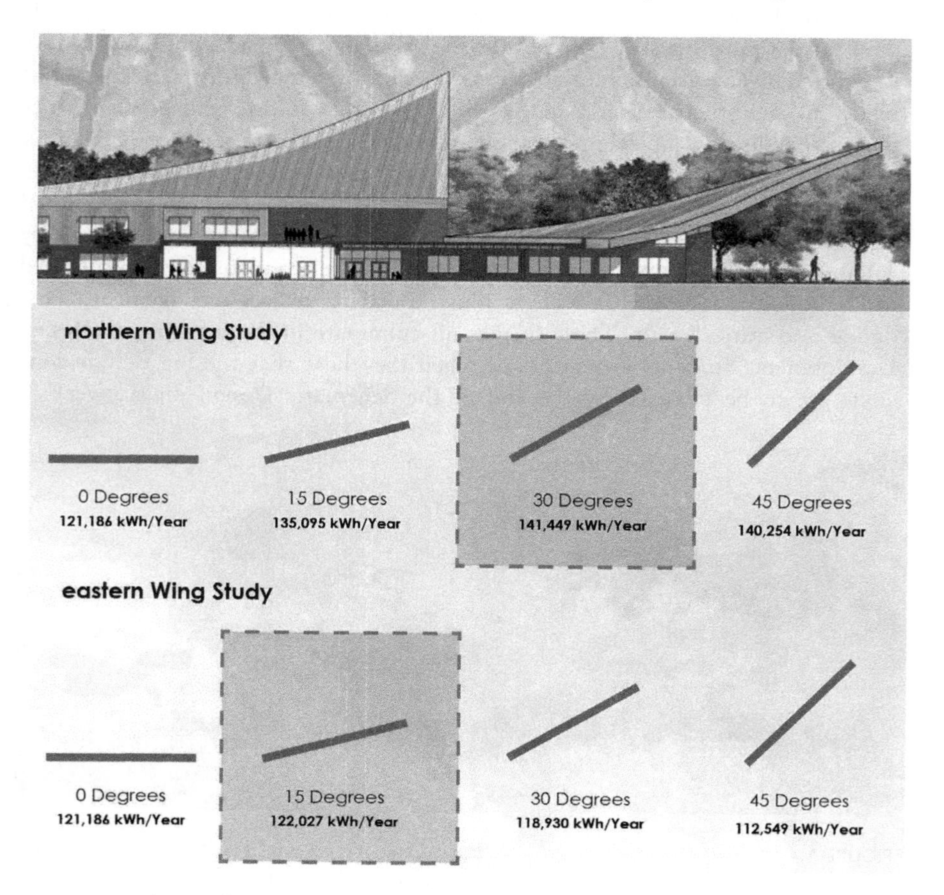

FIGURE 1.5 Photovoltaic roof study graphic using the PVWatts Calculator.
Source: Student work by Isabelle Cassidy and Schola Eburuoh

weeks of the semester. Students are tasked with telling their story. What did they do and why did they make the decisions they did? The story is to be told with two different graphic presentations, one for the in-person final jury and one for the AIA Top Ten for Students competition jury. In addition, the story needs to be detailed in a narrative format for both the in-person verbal presentation and the COTE competition required narratives. Each of these exercises requires iteration, and therefore students work with invited guests during these weeks to learn best practices to graphically display the work. Students also re-write their narratives one more time in response to instructor feedback on previous drafts. Finally, students work with 3M VAS software during these final weeks to understand what is likely to be seen pre-cognitively, i.e. what images are seen first, second, third, and how long each image is likely to be viewed. This is a way to ensure that the boards have been designed to highlight the graphic hierarchy that the students wish to communicate.

Analytics

This studio introduces a variety of software at each phase of design. Each semester, the software programs suggested for the course have evolved. Innovation is happening quickly and it is important that the students are exposed to

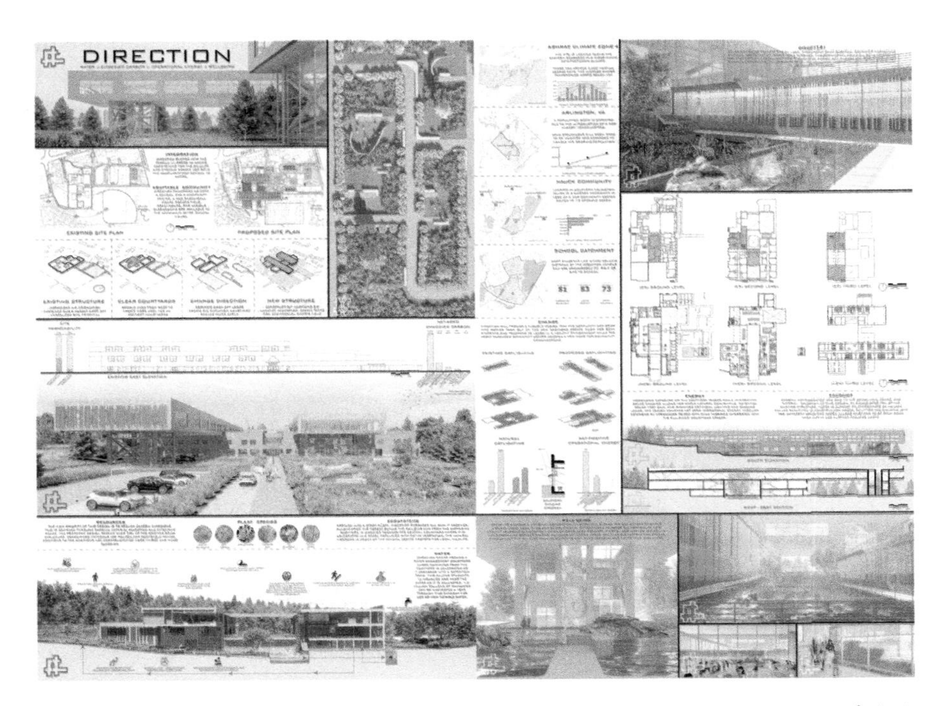

FIGURE 1.6 Final project boards for the Direction student group, winners of AIA COTE Top Ten for Students competition.
Source: Student work by Robert Conway and Morgan Weber

the most current tools available. Also, because the cost for software access changes frequently, software is selected each semester that is either free for students or available at an accessible cost for the student and/or the department to supplement.

As a result of all of the above, over the years, the course has required different combinations of Sefaira, EC3, Tally and cove.tool for operational energy, daylighting and embodied carbon decision-supporting calculations. Students can elect to use plug-ins for each of these with SketchUp, Rhino and Revit based on their familiarity with the programs and their point in the design process. In addition, supplementary websites have been recommended for student use throughout the semester such as: Cimate Consultant – to better understand the climate in which the site is located, PVWatts Calculator – to size PV arrays and understand their EUI offset contribution, construction carbon calculator – very early on in the design to understand the large-scale magnitude of their design decisions, and 3M VAS to understand the pre-cognitive effect of their designs on human health and well-being.

It is critical that two main points are emphasized during the semester. First, each software has its strengths and its weaknesses and thus, it is important to know how to use a wide variety of programs and which programs to use when. And second, each software has different levels of accuracy. Students are discouraged from presenting their analytics as exact. Instead, they are encouraged to use the analytics to inform and defend design decisions throughout the design process.

One final point, it has been shown that a key contributor to student success in this course is mastery of software as early as possible in the semester. Therefore, links to software tutorials are provided during the summer or winter break prior to the studio. Quizzes on each of the selected software are given in the first couple of weeks of the semester and the students need to keep re-taking them as required until they earn 80% or higher. For additional support throughout the semester, graduate assistants offer office hours.

Required Readings

Browning, William. "14 Patterns of Biophilic Design: Improving Health & Well-Being in the Built Environment." 2014. www.terrapinbrightgreen.com/reports/14-patterns/

Lechner, Norbert and Patricia Andrasik. *Heating, Cooling, Lighting: Design Methods for Architects*. 5th edition. John Wiley & Sons, Hoboken, New Jersey, 2021.

Living Building Challenge Program Manual (most recent version). www2.living-future.org/l/464132/2024-04-03/2bqwzrz

WELL Building Standard (most recent version). https://v2.wellcertified.com/en/wellv2/overview

Videos

Autodesk Sustainability Workshop: Building Design Strategy videos. www.youtube.com/@AutodeskEcoWorkshop

Net Zero Water – Bullitt Center Case Study by Sustainability Ambassadors. www.youtube.com/watch?v=EQ3WKgieSZ0
WELL: Healthier people through better buildings by International WELL Building Institute. www.youtube.com/watch?v=t5IfUe6j59w

Outreach Opportunities

This studio provides many opportunities for outreach both during the semester and after the completion of the course. During the semester, students visit the site, meet with the associated stakeholders and conduct fairly intensive community survey work. In addition, practicing architects and allied professionals visit the students often to share feedback. At the conclusion of the semester, students are encouraged to engage in community outreach efforts including local, regional and national competitions and conferences including the AIA National Conference, the AIA National Women's Leadership Summit and the DC Sustainability Summit. Students are also required to enter the AIA COTE Top Ten for Students competition. Undergraduate students who have taken this course often go on to attend graduate schools at other institutions including top programs at Harvard University, Virginia Tech, Clemson University, the University of Washington and the University of Oregon. In addition, students have also explored allied opportunities including the Peace Corps.

This course has also inspired outreach on a national level through the author's service on national AIA Committees such as the Higher Education Advisory Team (HEAT) and the Committee on the Environment Leadership Group (COTE LG). These advocacy groups have provided abundant opportunities for outreach via presentations at national academic conferences such as Association of Collegiate Schools of Architecture (ACSA) Annual Meetings, at national conference expo events which target student and emerging professional audiences, at open forum discussion tables focused on academic concerns and at White House listening sessions.

Perhaps the most applicable outreach; however, is that this course was the inspiration for this book, a compilation of 20 chapters, each describing unique and accessible pedagogies, united in the common goal of outreach.

Course Assessment

This course has been a highly popular course, both in terms of student enrollment and course evaluations. Student work from the studio has won the coveted AIA COTE Top Ten for Students competition as well as other regional design awards and has been selected for presentation at a variety of local, regional and national conferences.

Although this course meets most of the program and student criteria in the National Architecture Accrediting Board (NAAB) Conditions for Accreditation, it has been taught as an elective studio, and therefore, unfortunately, has not been considered formally as a contributing course toward the accreditation effort. This sheds light on a current curricular challenge: not all accredited programs require specific decarbonization course content in order to meet current NAAB accreditation requirements. One of the goals of this book is to encourage awareness and advocacy to revise these current criteria to include more specific language to this effect.

Looking Ahead

As stated above, it is critical that accreditation criteria be revised to include more stringent decarbonization language for architecture programs as soon as possible. To achieve that goal, it is essential that this curricular content become more widely accessible to reduce barriers to widespread adoption. Each chapter in this book offers unique, award-winning philosophies and pedagogies as a step toward this greater accessibility. It is the viewpoint of the author that the AIA's Framework for Design Excellence is a highly accessible tool that can be the foundation for architecture studios worldwide.

2

THE BUILDING AS A TEACHING TOOL FOR SCIENCE, TECHNOLOGY, ENGINEERING, ARCHITECTURE AND MATH

Erica Cochran Hameen and Nihar Pathak

Course Overview

Course Name

Architecture Design Studio: Praxis Studio 3

University / Location

Carnegie Mellon University / Pittsburgh, PA, USA

Targeted Student Level

- Undergraduate:
    - 4th year, Bachelor of Architecture, 5-year program

Course Learning Objectives

In this course, students will:

- Explore and demonstrate the necessary integration within the structural system, building envelope, environmental control systems and life safety systems while delivering measurable outcomes of building performance and occupant wellbeing as integral parts of the design process. Consulting engineers play active roles in the studio process, providing expertise and engaging in discussions resembling professional practice.
- Collaborate with several key stakeholder groups, focusing on education, equity, sustainability and planning.

DOI: 10.4324/9781032692562-4

- Create an engaging and exciting science, technology, engineering, architecture and math (STEAM)-focused elementary school through design exploration and in collaboration with architecture and engineering consultants, as well as key stakeholder groups. This school will be culturally sensitive, equitable, inclusive, energy-efficient, environmentally responsive and designed to foster interest in STEAM.

Course Learning Outcomes

Successful completion of this course will enable students to:

- Understand fundamental principles of structure, enclosure (envelope), environmental control systems and life safety systems.
- Develop the fundamental aspect(s) of the proposal through the measurable outcomes of building performance such as: daylighting, energy, natural ventilation and structural analysis.
- Develop clear, accurate and directed questions, and find thorough architectural training techniques and methods to study, evaluate and rework design thinking for an optimal response.
- Make technically clear drawings isolating the related elements that make up a comprehensive building.
- Understand the basic principles of the selection and application of envelope systems that address performance, aesthetics, moisture transfer, durability, energy use and material resources.
- Understand the basic principles and application of building services such as lighting, mechanical, plumbing, electrical, vertical transportation, security and fire protection.
- Demonstrate the skills associated with making integrated decisions across multiple systems: including identifying challenges, determining evaluative criteria for making decisions, analyzing potential solutions and determining the effectiveness of its implementation.

Course Methodology

Philosophy

The teaching philosophy was built on the foundation of heritage, education, professional experience and research experience to be incorporated together to drive the designs towards sustainable designs that also curtail the impacts on the environment. This foundational background fuels the motivation to provide positive, meaningful and fruitful learning environments for all students. The teaching methodology merges design, architecture, art, technical and critical thinking approaches and aims to create a learning environment where everyone can build and capitalize on their personal strengths, gain and retain knowledge/

the subject matter, and make personal connections with the importance/criticality of the topic and how it relates to their past, present and future.

The teaching philosophy and methods are student-focused and built upon the understanding that all people learn best, utilizing different teaching methods and at different paces. Additionally, while a student may be a fast learner in one topic area, that same student may learn at a different pace in another topic. The student-focused teaching methodology adjusts to the course structure and student educational level and adapts to the needs and strengths of the student population of each course.

The teaching philosophy is rooted in working with students and people from all backgrounds and demographics, sharing knowledge and helping students understand the course objectives and the importance of the subject matter. Critical to the teaching philosophy is bringing the building occupant to the center of all discussions related to sustainability and climate change. The teaching philosophy's proposition is that we must bring a face to sustainability. It is not enough to measure environmental improvements in kWh, BTU and carbon emission reductions, we must connect those metrics and numbers to the building occupants and identify how people, families and communities will be impacted by those metrics.

Definitions

Carbon neutrality – Addressing sustainability principles through design to reduce carbon emissions through improving indoor environmental quality, energy efficiency, and the use of materials with lower embodied carbon content.

Building as a teaching tool – The process of exposing building systems and utilizing interactive technology in buildings as a strategy to teach occupants about architecture, sustainability, engineering, structure, and mechanical, electrical, plumbing and technological systems as a means to facilitate greater learning and a deeper understanding of architecture and engineering.

Community engagement – Working with the community to provide them with quality spaces by providing meaningful engagement between the community and the design team (architects and engineers) during the design process by providing hands-on experiences and utilizing common language.

Design justice – The process of utilizing design to address structural inequalities in marginalized and disenfranchised communities and providing building occupants with design strategies that promote occupant well-being, health and productivity.

Energy modeling – Computer generated calculations to highlight the impacts of design strategies used to predict energy usage.

Equitable sustainability – The process of evaluating, understanding, monitoring and providing holistic solutions to unhealthy environmental and physical conditions that vulnerable populations are subjected to because of historical and continuing inequities and providing efficient holistic solutions that minimizes the environmental impact and enhances occupant well-being, occupant comfort and occupant productivity.

Indoor environmental quality (IEQ) – The five metrics utilized to define and evaluate the indoor environment; they include air quality, visual quality, thermal quality, acoustical quality and spatial quality.

Simulations – Computer aided design alternatives used to identify the potential impacts of design strategies and predict the outcome of performance under set conditions. These include categories such as daylighting, glare, energy and radiation analysis.

STEAM – Science, technology, engineering, architecture and math.

STEM – Science, technology, engineering and math.

Implementation

General

A well-rounded education provides tremendous opportunities. School facilities are not just spaces for education, they are places where people can learn, enhance and develop their understanding of the world, make friends, play, explore, create, grow, secure a better income, develop critical thinking and use their imagination to make the impossible possible. In the United States, students in elementary and high schools spend, on average, 1,400 hours in school buildings every year, engaging in learning, playing, eating and interacting with one another.

Students are given a semester-long project to design an elementary school named, The STEAM Center of Learning where the school building serves as the teaching tool for STEAM education.

Throughout the semester, design solutions are measured utilizing qualitative and quantitative methods.

Qualitative methods are subjective and include personal and subjective assessments to describe an object or solution. They can be recorded utilizing multiple descriptive forms including photographs, personal verbal and non-verbal expression, structured and unstructured inquiry, and collecting and analyzing non numerical data, opinions, experiences and ideas. Quantitative methods are objective and include assessment of numerical data. They include the assessment of objects and solutions that can be measured numerically. Throughout the semester students are asked to demonstrate, document, provide and present their design solutions, and participate in discussions of other solutions and techniques using both qualitative and quantitative methods.

Based on the number of students in the class, students are divided into teams of 3–4. One person per team shall serve as the project manager for structure, enclosure, mechanical, electrical and plumbing (MEP) and documentation. The team project managers are ultimately responsible for making team assignments and producing the deliverables throughout the semester.

The studio is part of an integrated studio led by Professor Gerard Damiani and with Professor Stephen Lee, where the syllabus is developed in conjunction with them to prioritize the students' learning and access to knowledge and skills that would ensure their progress.

Series of conversations

Each week, the class includes time for conversations related to the interconnectivity of architecture, equitable sustainability, community engagement and the role of the architect. Some of the topics covered in this course may be difficult to discuss. The students are informed that we acknowledge that there will be differences in opinions and beliefs. We request that the class respect all opinions and engage the topics with empathy, especially as we discuss energy and built environment inequalities that are especially prevalent in communities with large populations of Black, Brown and Indigenous people and issues that disproportionately and negatively impact Black, Brown, Indigenous and low-income communities. We urge the class to have the courage and spirit to speak out against injustices, to use their abilities and intellect to challenge the status quo and to help identify new and innovative solutions that will provide a more energy efficient, just, healthy, equitable and sustainable world.

Assignments

The studio is divided into four major pin-up reviews that encourage the students to develop their projects through a storyline. The class is divided into groups of 3–4 students per group, thus enabling a strong foundation for collaborative work and preparing them for a professional practice office environment. Each review includes a multi-step submission process which allows the students to distribute their work and manage their time more appropriately.

Review 1 includes a series of design introductory activities. The students are introduced to their project to provide a general idea of the project scope and program while still providing opportunities for design and program flexibility. The initial sessions start with a presentation to the students of architectural building case studies of inspiring architectural projects. Students are then required to conduct individual research to identify additional case studies of building types that provide learning in a fun environment. Furthermore, they are required to document memories and ideas from their past experiences and "fun spaces and built environments' they reminisce from their youth. The students then head out for a site visit where they have a "guided walk in the neighborhood' to understand the site context, interact with the community members and even just to grab a meal while they are there for a deeper sense of connection with the community, their culture and their experiences. Such actions of community engagement lead to a robust understanding of the needs of the community, thus helping the students plan for a better suited design.

After their site visit, the students are asked to present their case studies and form groups according to their preferences. During these presentations, there is an exchange of ideas as well as development of the concepts that they think would enhance the experience of the community. The groups then proceed to perform further research about the building type, the community, the neighborhood, its history and environmental factors thus leading to address their proposals with design justice. This is a collaborative work among the team members which enhances their understanding of the project and the site and to seek varying perspectives about the same. The teams then proceed to write a concept of their intent for the project in the form of words. They have the freedom to express themselves in the form of keywords, short narratives, or simple sketches. These concepts are then presented and further developed with more inputs from their peers.

The teams are then provided programs for the projects which include design areas for some spaces but not all. The students circle back to their research to identify how big certain areas need to be according to the community's needs. Once the program is developed (Figure 2.1), the teams engage in putting together conceptual design options in lieu of the performed site analysis that addresses the environmental factors using building as a teaching tool. They start developing block models and conceptual sketches at this time. The teams, with the help of the studio instructors, work towards finalizing one of the concepts and it is then included in the first review session. The teams each receive feedback from various faculty about the concept being developed and the intention towards the project. In the case of the chosen concept not being selected, the faculty works with the team to revisit their other concepts to pick another one or combinations of the others to develop a new one. By the end of the first review, the final concepts are chosen for each of the teams and they proceed to work on these for the rest of the semester.

Review 2 includes further development of the design that is graded on 7 defined criteria. These criteria include enclosure, climate & energy (Figures 2.3 and 2.4), MEP (Figure 2.5), structural, design process, design integration and presentation (Figure 2.2).

Assignments 1 and 2 are utilized to identify the mid-term grades for the students and to help them understand the grading criteria. Furthermore, it helps the students develop the designs in a meticulous way. During these reviews, the teams receive feedback from various faculty members, the clients, as well as industry professionals leading them to improve on the design.

Reviews 3 and 4 follow a similar framework to Review 2, where the development is based on the 7 criteria (Figures 2.6 and 2.7). The students are also provided with a detailed list of 27 points that they are expected to cover through all the reviews, thus clearly establishing a clear expectation and guideline for the students to follow. The teams partake in understanding how the MEP designs would work for their design and generate models along with their

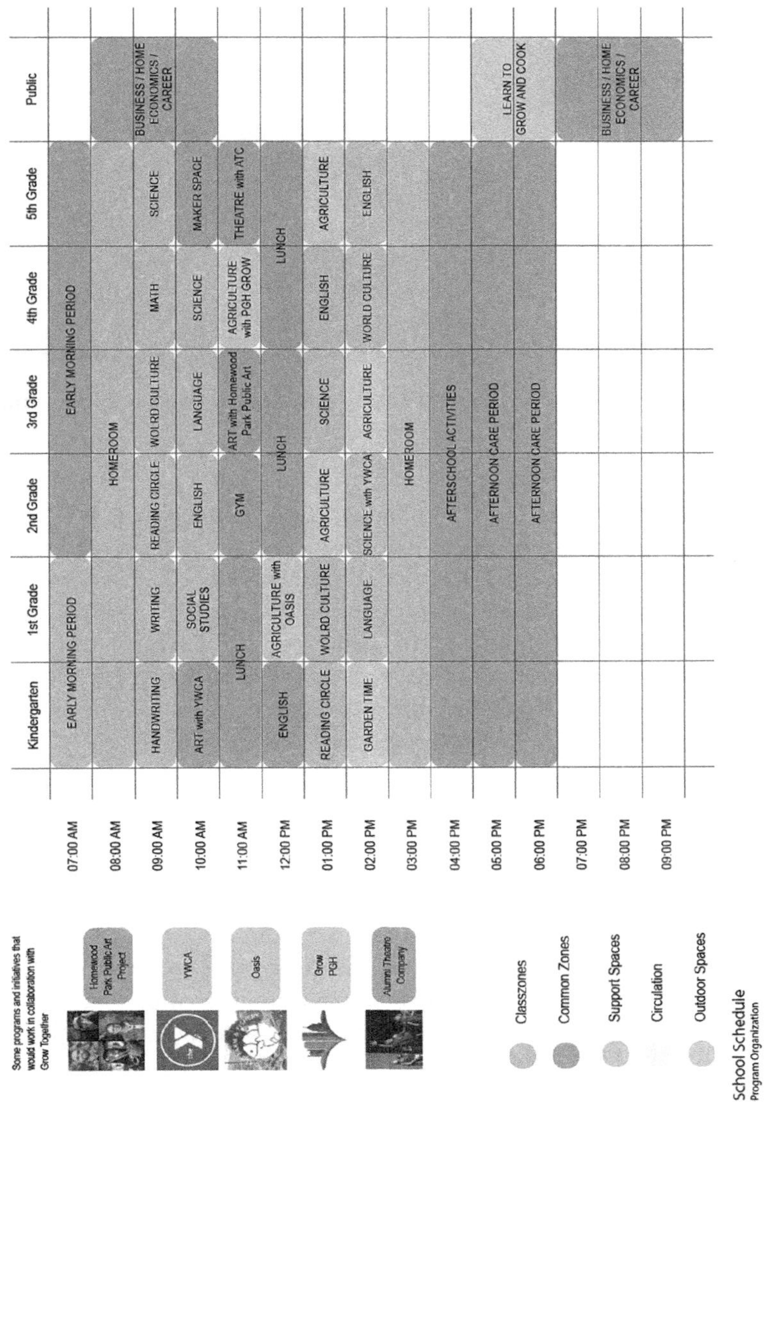

Some programs and initiatives that would work in collaboration with Grow Together

Homewood Park Public Art Project

YWCA

Oasis

Grow PGH

Alumni Theatre Company

	Kindergarten	1st Grade	2nd Grade	3rd Grade	4th Grade	5th Grade	Public
07:00 AM	EARLY MORNING PERIOD		EARLY MORNING PERIOD				
08:00 AM			HOMEROOM				BUSINESS / HOME ECONOMICS / CAREER
09:00 AM	HANDWRITING	WRITING	READING CIRCLE	WOLRD CULTURE	MATH	SCIENCE	
10:00 AM	ART with YWCA	SOCIAL STUDIES	ENGLISH	LANGUAGE	SCIENCE	MAKER SPACE	
11:00 AM	LUNCH		GYM	ART with Homewood Park Public Art	AGRICULTURE with PGH GROW	THEATRE with ATC	
12:00 PM	ENGLISH	AGRICULTURE with OASIS	LUNCH		LUNCH		
01:00 PM	READING CIRCLE	WOLRD CULTURE	AGRICULTURE	SCIENCE	ENGLISH	AGRICULTURE	
02:00 PM	GARDEN TIME	LANGUAGE	SCIENCE with YWCA	AGRICULTURE	WORLD CULTURE	ENGLISH	
03:00 PM			HOMEROOM				
04:00 PM			AFTERSCHOOL ACTIVITIES				
05:00 PM			AFTERNOON CARE PERIOD				LEARN TO GROW AND COOK
06:00 PM			AFTERNOON CARE PERIOD				
07:00 PM							BUSINESS / HOME ECONOMICS / CAREER
08:00 PM							
09:00 PM							

Classzones

Common Zones

Support Spaces

Circulation

Outdoor Spaces

School Schedule
Program Organization

FIGURE 2.1 Program developed by "Together we grow" team for the first review.

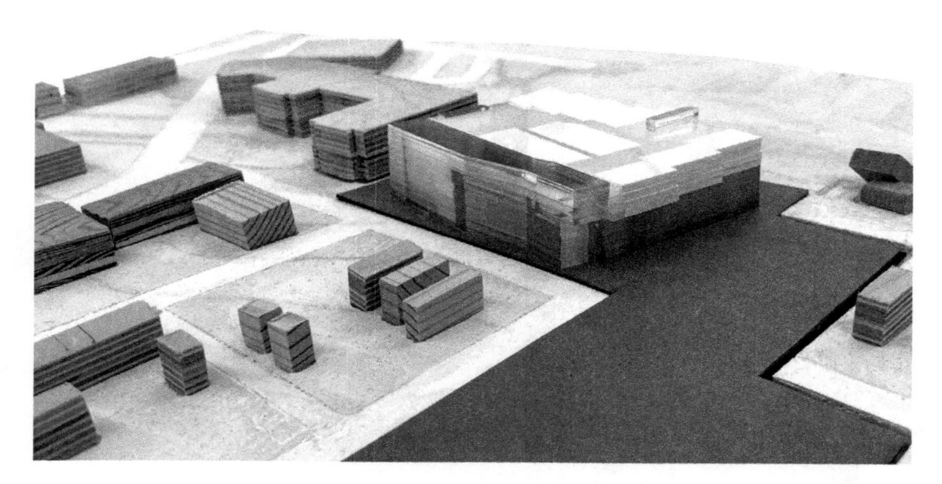

FIGURE 2.2 Conceptual models by the "Water" team using wooden blocks and acrylic sheets.

MEP layouts as seen in Figure 2.3. Using building as a teaching tool, the teams strive to integrate multiple strategies to enhance their designs from a building science perspective and designing for equitable sustainability. Utilizing simulations (Figures 2.3 and 2.4) and energy modeling to compare and analyze their designs and proposed design solutions, the teams attain a design that is more efficient in terms of energy consumption, reducing the carbon footprint and providing a higher comfort level for the occupants in terms of the indoor environmental quality. Review 4 serves as the final pin-up review.

Analytics

As part of the learning objectives, students are required to use simulation software to support their sustainable design concepts. To accomplish this goal, they use building energy and performance simulation software that they learned in previous classes. Throughout the semester, the students are also exposed to new simulation software and encouraged to experiment with the simulation software.

To increase learning at the undergraduate and graduate levels, students in the studio class are paired with PhD students in the Building Performance and Diagnostics program. This provides benefits to the undergraduate students since they are able to meet with simulation experts and have someone other than their studio professor provide feedback on their simulation results. It also is a learning opportunity for the PhD students, many of whom may become university professors. The PhD students are provided with opportunities to share their knowledge and teach the software with the studio professor.

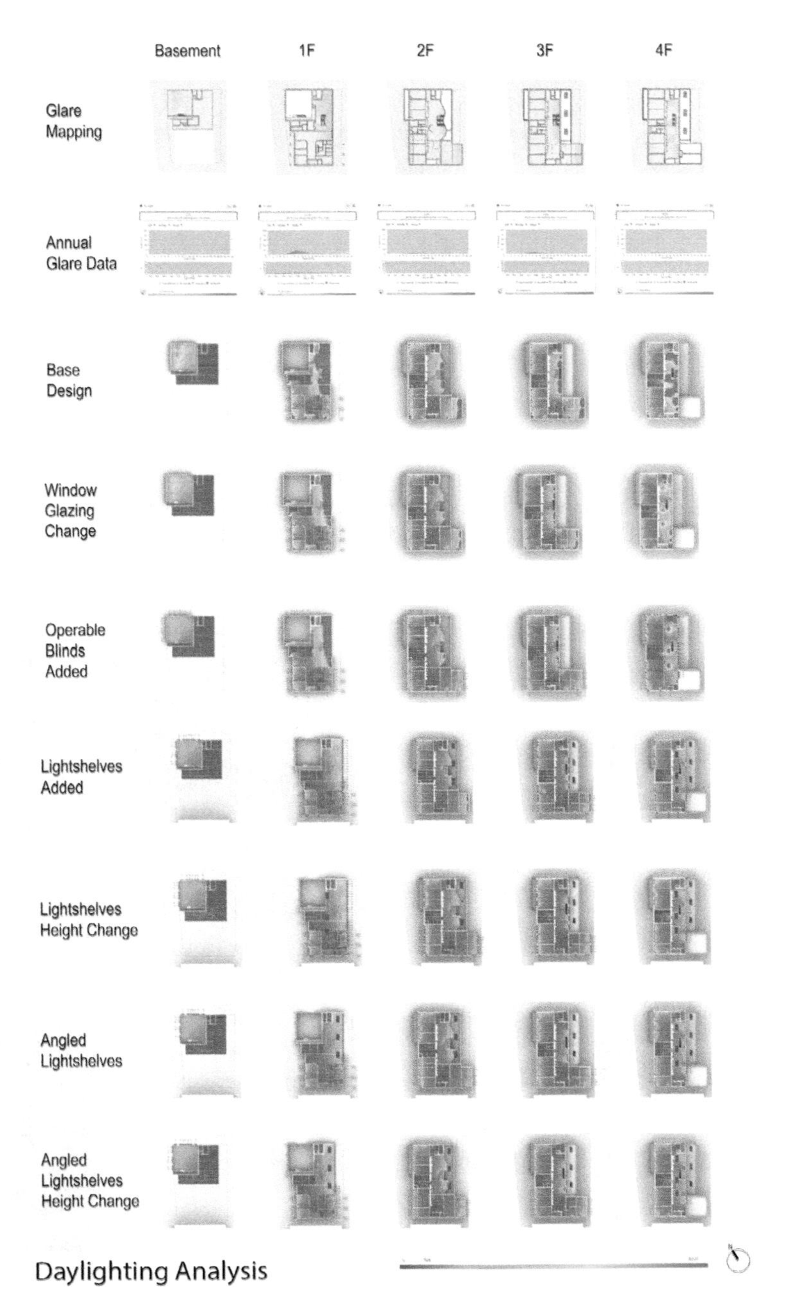

FIGURE 2.3 Daylighting analysis conducted by the "Together we grow" team using Rhino and Climate Studio.

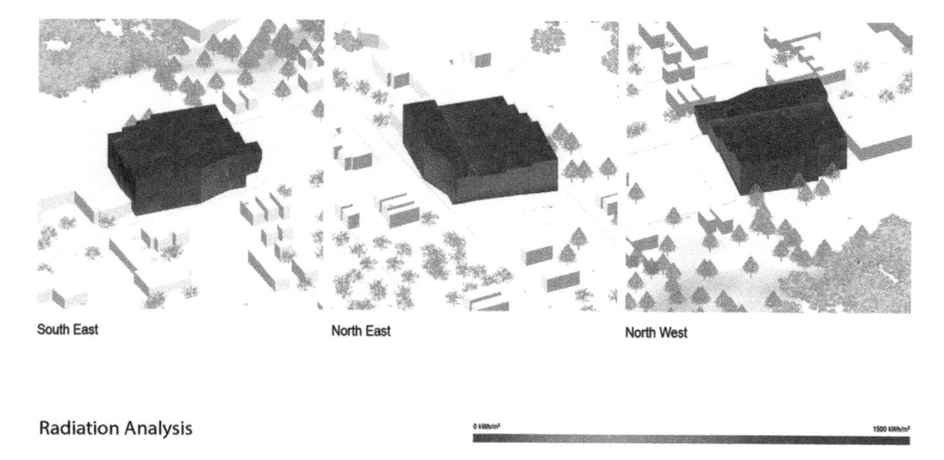

FIGURE 2.4 Radiation analysis by the "Water" team at the concept development stage using Rhino and Climate Studio.

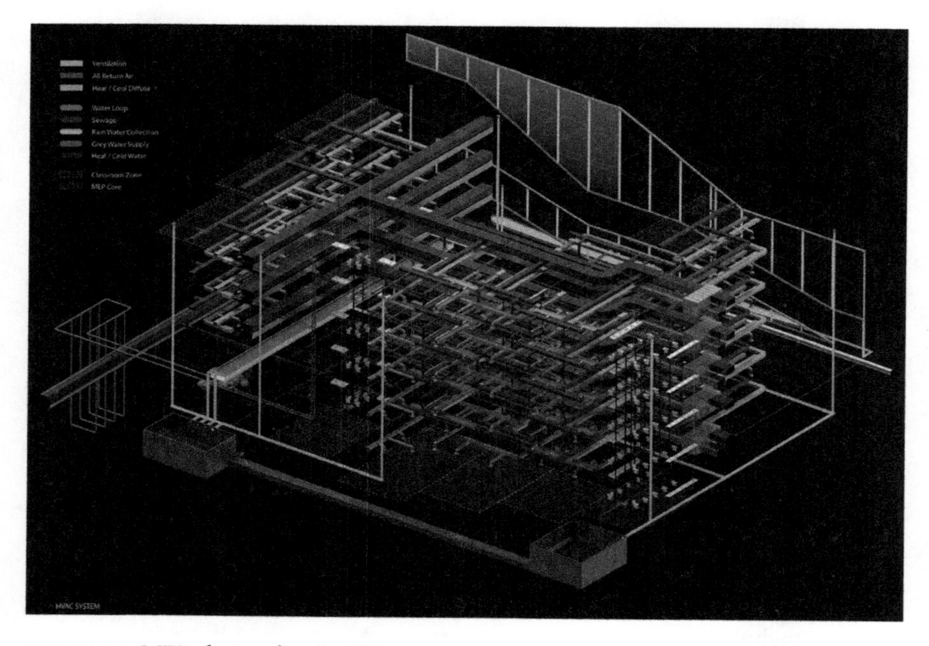

FIGURE 2.5 MEP design for the "Water" team.

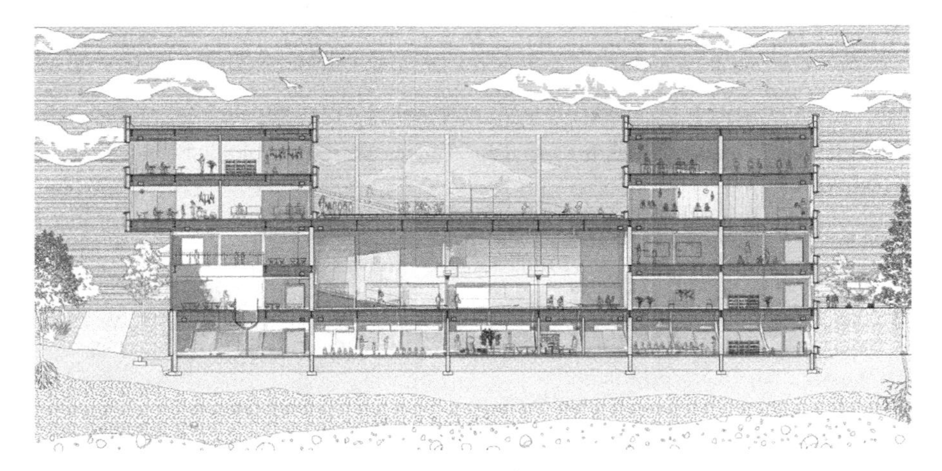

FIGURE 2.6 Section for "Water" team.

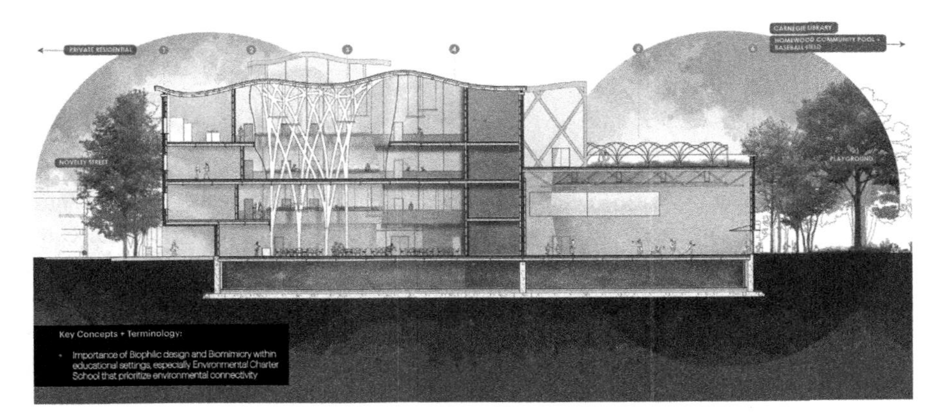

FIGURE 2.7 Sectional drawing of Environmental Charter School in Homewood by Christina X Brown.

Students are encouraged to run multiple energy, air, lighting and thermal simulations and examine how the simulation results change when they make architectural design changes to the building.

Required Readings

A Union of Professionals. (2006). Building Minds, Minding Buildings: Turning Crumbling Schools into Environments for Learning. See https://files.eric.ed.gov/fulltext/ED497873.pdf

California Energy Commission. (2008). AB 2160 Green Building Report for Submission to The Governor's Green Action Team. State of California Energy Commission.

Cochran Hameen, E., Ken-Opurum, B., & Son, Y. J. (2020, May). Protocol for Post Occupancy Evaluation in Schools to Improve Indoor Environmental Quality and Energy Efficiency. *Sustainability*, 12(9).
Filardo, M. (2008). Good Buildings, Better Schools: An Economic Stimulus Opportunity with Long-Term Benefits. Briefing Paper #216. Washington DC: Economic Policy Institute.
Kats, G. (2006). *Greening America's Schools: Cost and Benefits*. See www.usgbc.org/resources/greening-america039s-schools-costs-and-benefits
Soroye, K. L., & Soroye, K. L. (2010). Building a Business to Close the Efficiency Gap: The Swedish. *Energy Efficiency*, 3, 237–256.
Taylor, R. P., Govindarajalu, C., Levin, J., Meyer, A. S., & Ward, W. A. (2008). *Financing Energy Efficiency: Lessons from Brazil, China, India, and Beyond*. Washington, DC: World Bank.
Valentine, M., & Vargas, D. (2007). *Green Design and Practice in the Development of Buildings and Neighborhoods*. Arabella Philanthropic Advisors. Funders' Network.
Wiley, J. A., Benefield, J. D., & Johnson, K. H. (2010, July 30). Green Design and the Market for Commercial. *Journal of Real Estate Finance and Economics*, 41, 228–243.
Xu, T., & Piette, M. A. (2004, July 23). *Baseline Information Development for Energy Smart Schools – Applied Research, Field Testing and Technology Integration*. See https://eta-publications.lbl.gov/publications/baseline-information-development

Outreach Opportunities

Community Connections

The students engage with key community stakeholders with expertise in educational buildings and K-12 learning. These stakeholders include Pittsburgh community leaders, officials from Allegheny County and the Pittsburgh Government, as well as leaders from the Pittsburgh Architecture Learning Network (ALN). The ALN encompasses leadership from the ACE High School Mentorship Program, Assemble, CMU School of Architecture, the Carnegie Museum of Art, Fallingwater, Pittsburgh History and Landmarks Foundation, the University of Pittsburgh Architectural Studies Program, Chatham University Interior Architecture, the National Organization of Minority Architects (NOMA) PGH, and the Young Preservationists Association of Pittsburgh.

Throughout the semester it is important for students to have many opportunities to engage with multiple community stakeholder groups to gain a deeper understanding of the community. The goal is to introduce students to participatory design concepts and help students understand the complex issues and challenges faced by communities from different backgrounds with various levels of resources.

Professional Connections

Each student team has the opportunity to speak with professional consultants who provide professional advice in their specialty areas. The goal is to aid the students in not only developing architectural drawings that are clear, sustainable and technically correct and but to understand the role consultants play in bringing a project to fruition, and the importance of collaboration among a diverse group of professional expertise.

Enclosure Consultant

Gensler (www.gensler.com/)

- Charles Coltharp, Associate | Technical Director
- Prerana Paliwal, Technical Designer | Co-founder MAPS

MEP Consultant

Smith Group (www.smithgroup.com/)

- Megan Campbell, Mechanical Engineer
- Colin Pocock, Mechanical Engineer

Note that, for Smith Group, the two engineers are junior level, and when needed we also have consultation from two senior consultants:

- Katrina Kelly-Pitou, Principal, Energy Systems Strategist
- Ling Almoubayyed, Project Architect

Structural Consultant

Carnegie Mellon University

- Dr. Juney Lee, Assistant Professor

Collaborative Research Connections

Students in the studio will participate in an interactive workshop focused on the development of an interactive toolkit for designing K-12 sustainable schools. The workshop is part of a PhD dissertation titled, "Safe and Healthy Schools – Synergistic Design Actions to Improve School Buildings." The workshop will be led by Annie Ranttila, AIA, PhD Candidate, CMU, Coordinating Architect at Chicago Department of Aviation.

Field Trips

- Assemble, a community space for Art + Technology, Pittsburgh, PA, USA
- The Children's Museum and the Museum Lab, Pittsburgh, PA, USA

- Phipps Center for Sustainable Landscapes, the Exhibit Staging Center, and the Nature Lab, Pittsburgh, PA, USA

Course Assessment

Throughout the semester, students have monthly meetings with the studio professor and the teaching assistants to review their progress in each of the defined criteria metrics. The well-defined criteria and frequency of meetings are essential to ensure student success and comprehension of the learning objectives. The meetings and clear feedback on each criterion also provide students with specific recommendations on areas where they need to provide additional focus or where the faculty can provide additional instruction to ensure student success. The meetings are a two-way conversation where students also provide feedback on the teaching methodology and which methods are most effective in their learning process. This two-way conversation provides an opportunity for both students and faculty to make changes or modifications throughout the semester. Utilizing this methodology, at the end of the semester, all students had successful design projects and each student said they received the grade they were expecting. The students also expressed appreciation for the clarity and frequency of the feedback.

In addition to the faculty and teaching assistant feedback, students receive feedback from other department faculty and the consultants at each of the four reviews. Additionally, students receive feedback during the midterm and final reviews from architects and allied professionals.

At the conclusion of the semester, each student is provided access to a formal online anonymous faculty course evaluation survey conducted by Carnegie Mellon University. Additionally, students are invited to debrief meetings with the studio faculty to discuss their progress throughout the semester. During these sessions, students also provide suggestions for the teaching methodology and semester structure that could help future students of the course. Finally, students receive feedback on modifications, if any are needed, that they should make to enhance the quality of their portfolio.

The course is also assessed using National Architectural Accrediting Board (NAAB) Criteria, SC.6 – Building Integration: How the program ensures that students develop the ability to make design decisions within architectural projects while demonstrating integration of building envelope systems and assemblies, structural systems, environmental control systems, life safety systems and the measurable outcomes of building performance.

Looking Ahead

The teaching philosophy ensures that the students engage with more hands-on experience rather than the traditional ways. This is possible based on the professor's connections and involvement with community engagement and development built upon the foundation of heritage, education, professional experience and research experience. The concepts and skill sets developed in this course have multiple avenues that can be applied in the future. This includes, but is not limited to, graduate study research inclining towards indoor environmental quality and energy efficiency that holds the occupants' health and comfort as its priority. There is also a possibility to continue working with communities to enhance their building experience as well as engagement with STEM education. The course showcases its collaborative nature and engagement with varying audiences. Post publication, this will prove to have opportunities to team up with other universities, professional practices and communities to embrace engagement and collaborate, enabling the future generation of young architecture professionals to seek more than just theoretical studio-based projects.

3

INTEGRATED AND INTERDISCIPLINARY DESIGN EDUCATION FOR NET-ZERO ENERGY BUILDINGS

Edoarda Corradi Dell'Acqua

Course Overview

Course Names

Net-Zero Energy Building Design I & II

University / Location

Illinois Institute of Technology / Chicago, IL, USA

Targeted Student Level

Net-Zero Energy Building Design I & II are required courses for graduate students enrolled in the Master of Engineering in Architectural Engineering and the Master in High-Performance Buildings programs. Additionally, they are open to undergraduate students across all majors through Illinois Tech's Interprofessional Projects (IPRO) program and graduate students from other programs. The IPRO program is a required undergraduate academic program that brings together students from diverse majors to explore innovative solutions to real-world challenges.[1]

- Undergraduate
  - 3rd or 4th year, any major, IPRO program (Kaplan Family Institute for Innovation and Tech Entrepreneurship)
- Graduate
  - 1st or 2nd year, Master of Engineering in Architectural Engineering, Department of Civil, Architectural and Environmental Engineering (CAEE)[2]

DOI: 10.4324/9781032692562-5

- 1st or 2nd year, Master of High-Performance Buildings, jointly offered by the Department of Civil, Architectural, and Environmental Engineering (CAEE) and the College of Architecture (CoA)[3]

Course Learning Objectives

In these courses, students will:

- Follow the iterative process of integrated building design to complete plans for a low carbon and net-zero energy capable building.
- Develop designs and produce comprehensive documentation for a high-performance residential, mixed-use, or commercial building.
- Research and implement innovative net-zero energy designs, technologies, and systems.

Course Learning Outcomes

Successful completion of these courses will enable students to:

- Work on multidisciplinary teams that reflect real-life experience in the architecture, engineering, and construction (AEC) industries.
- Critically analyze building design elements for their impacts on energy use, carbon emissions, health, comfort, aesthetics, and costs.
- Properly size on-site renewable energy systems.

Course Methodology

Philosophy

Designing a high-performance building requires close collaboration across various disciplines and areas of expertise. Net-Zero Energy Building Design I & II – a sequence of two courses centered on the US Department of Energy (DOE) Solar Decathlon Design Challenge[4] – aim to promote and strengthen collaboration between Illinois Institute of Technology's architecture and engineering students, narrowing the gap between the two disciplines. These courses are core components in two graduate degree programs open to students in the Department of Civil, Architectural, and Environmental Engineering (CAEE) and the College of Architecture (CoA): The Master of Engineering in Architectural Engineering and the Master of High-Performance Buildings. The latter is the first degree jointly offered by Armour College of Engineering and the College of Architecture. Furthermore, since spring 2022, Illinois Tech's IPRO program has expanded access to these courses to undergraduate students from all academic majors.

The primary objective of these courses is for students to develop designs and produce comprehensive documentation for a zero energy building. Throughout this process, students engage in interdisciplinary teamwork. Cultivating students' ability to collaborate within multidisciplinary teams and effectively communicate across disciplinary boundaries is a significant focus of these courses. By listening to each other's diverse perspectives, students learn to prioritize objectives and reach compromises that enhance the project's overall outcome and lead to a balanced design addressing programmatic and spatial needs while meeting the energy performance goals. Students are encouraged to examine the broader context, including social and historical factors, equity considerations, cost-effectiveness, health impacts, feasibility and resilience to climate change.

Collaboration and exchange of ideas are fundamental aspects of the class and occur at every level. Faculty members from various departments contribute to the course through guest lectures and offering design advice. Moreover, industry professionals actively participate in project reviews and lectures, providing students with practical insights into real-world challenges.

Promoting and leveraging collaboration across different colleges and fields while strengthening ties with the industry broadens students' horizons and fosters a holistic design approach.

Definitions

Carbon neutrality (built environment) – The goal to balance the building sector's impact on greenhouse gas (GHG) emissions. It involves reducing the project's embodied carbon, increasing energy efficiency, and replacing fossil fuels with renewable energy sources.

Holistic design – A comprehensive approach that considers all aspects of the design, encompassing the user's needs and the project's broader context, including health and well-being, social, environmental, economic, cultural, and historical factors.

Integrated design – An approach that focuses on the relationships and interdependencies between various parts of a design. It ensures that every part serves the same overarching goal and supports other elements, promoting cohesion, synergy, and efficiency across the project.

Interdisciplinary design – A collaborative process that brings together expertise, methods, and perspectives from diverse disciplines to develop innovative and comprehensive solutions to complex problems.

Zero energy building (ZEB) – "A zero energy building (ZEB) produces enough renewable energy to meet its own annual energy consumption requirements, thereby reducing the use of non-renewable energy in the building sector."[5]

Implementation

General

Course Development

Illinois Institute of Technology's involvement in the Solar Decathlon Design Challenge, formerly the Race to Zero competition, began as an extracurricular activity in 2016. It stemmed from the enthusiasm of a group of architecture and engineering students who expressed interest in participating in the competition.[6] This experience highlighted the educational value offered by the Solar Decathlon and the importance of incorporating and formalizing high-performance building design courses in the architectural engineering curriculum.

Following this first participation, the Department of Civil, Architectural, and Environmental Engineering (CAEE) developed a two-course sequence offered yearly in the fall and spring semesters and built around the Design Challenge. These courses initially started as optional mixed-level undergraduate/graduate courses open to engineering and architecture students and developed into mandatory graduate courses for two of the master's programs. Furthermore, the recent integration of this sequence into Illinois Tech's IPRO program provides a pathway for undergraduate students from various disciplines to engage in the competition.

Including this sequence in the IPRO program has increased enrollment and participation from architecture students. Initially offered by Amour College of Engineering, enrollment in these courses was skewed towards engineering, with a ratio of approximately 2:1 engineering to architecture students. However, through IPRO, this ratio has balanced out, with architecture and engineering students each comprising around 50% of the course enrollment. Additionally, students from various disciplines, such as Electrical and Computer Engineering, Chemical Sciences, Biological Sciences, Information Technology & Management, and Physics, have joined the teams.

Net-Zero Energy Building Design I & II have effectively bridged two colleges and one institute at Illinois Institute of Technology, including Armour College of Engineering, the College of Architecture, and the Ed Kaplan Family Institute for Innovation and Tech Entrepreneurship.

Course Structure

The Solar Decathlon Design Challenge timeline follows the academic calendar, with the semifinal and final events occurring in the spring semester. Net-Zero Energy Building Design I & II meet once a week over two semesters. Articulating these courses over two semesters has several advantages. Most importantly, it provides enough time for students to thoroughly understand and process the material, recognizing how various parts of the project relate to each other.

Moreover, the extended timeline allows students from different colleges to get to know each other, fostering cohesive teamwork as teams gradually come together.

The first part of the sequence is dedicated to gathering the necessary background knowledge to design a zero energy building. This involves students completing the DOE's Solar Decathlon Building Science Education Series[7] to understand building science principles and their application to design. During the fall, students also familiarize themselves with various software tools for project design and analysis. Moreover, the fall semester focuses on activities such as team formation, selection of projects and sites, and developing connections with industry and community partners.

Vertical River,[8] presented at the 2020 Design Challenge, offers an example of collaboration between students and industry. During the fall semester, students established a connection with the architecture, interior design, and planning firm SCB, which subsequently became an industry partner for this project. Inspired by SCB's urban vision for the River District along the Chicago River's North Branch, the team designed a 15-story office building within SCB's master plan. *Vertical River* won first place in the Office Building Division.

With no major competition deadlines during the fall, students can dedicate their time to researching innovative technologies and approaches to the design of high-performance and resilient buildings. For the 2024 Design Challenge, one team explored the application of a buoyant foundation for a single-family home in a flood-prone Miami area.[9] Buoyant foundations, a type of amphibious foundation, are an innovative strategy that enables buildings to float during a flood event – the project aimed to provide resilient shelter for low-income residents who cannot evacuate during a storm.

Over the years, teams have explored different project types, ranging from small-scale single-family homes to commercial buildings, including new construction and retrofit projects of existing buildings.

FIGURE 3.1 *Vertical River* office building along the Chicago Riverfront.

FIGURE 3.2 *The Coastal Refuge* single-family home during a flooding event.

In the spring semester, teams finalize the designs. In addition to further developing the project's technical aspects, the focus includes crafting a cohesive narrative and refining graphic and verbal presentations.

Industry and Community Partnerships

Throughout the two semesters, industry partners from diverse disciplines within the Architecture, Engineering, and Construction (AEC) sector offer continuous support to student teams by providing technical guidance.[10] This support is facilitated through two formal presentations each semester: a midterm and a final presentation. By the end of the course sequence, teams present their work in class four times to design professionals and at least once to experts from various sectors during Innovation Day, a student project exhibit organized at the end of the fall and spring semesters by the IPRO program. The purpose of these meetings is for students to collect feedback on the proposed designs. Additionally, they offer students the opportunity to practice their presentation skills for clarity and conciseness and their ability to answer questions effectively. Former students return to the classroom as industry partners, offering advice to the teams based on their experiences as students and professionals.

Teams have also collaborated with community partners, addressing real-life projects. For instance, in 2021, students connected with the Phyllis Wheatley Home Group and Preservation Chicago[11] to explore solutions for preserving Chicago's last-standing Phyllis Wheatley Home on the South Side. Designed by the renowned architect Frederick B. Townsend and constructed in 1896, this historical mansion is at risk of demolition due to its deteriorating structural conditions and code violations. For over 50 years, the Phyllis Wheatley Home has served as a haven for black women who arrived in Chicago after the Great Migration and Preservation Chicago has been actively advocating for its preservation.[12] The team's proposal reimagined the historic structure as a place serving the community, giving it a renewed purpose, and enhancing its overall energy performance.[13]

In 2023, two Illinois Tech teams collaborated closely with Sunshine Gospel Ministries,[14] a not-for-profit organization on Chicago's South Side. Through its Housing Equity Initiative (HEI),[15] Sunshine Gospel Ministries aims to promote homeownership in a community that historically suffered from disinvestment and marginalization. One team focused on retrofitting an existing single-family home.[16] Meanwhile, the second team developed a design for a mixed-use multifamily building named *The Growing Haven*[17] on a vacant city lot. Both project sites were classified as disadvantaged by the Justice40 initiative.[18] Principles of affordability and equity guided both projects.

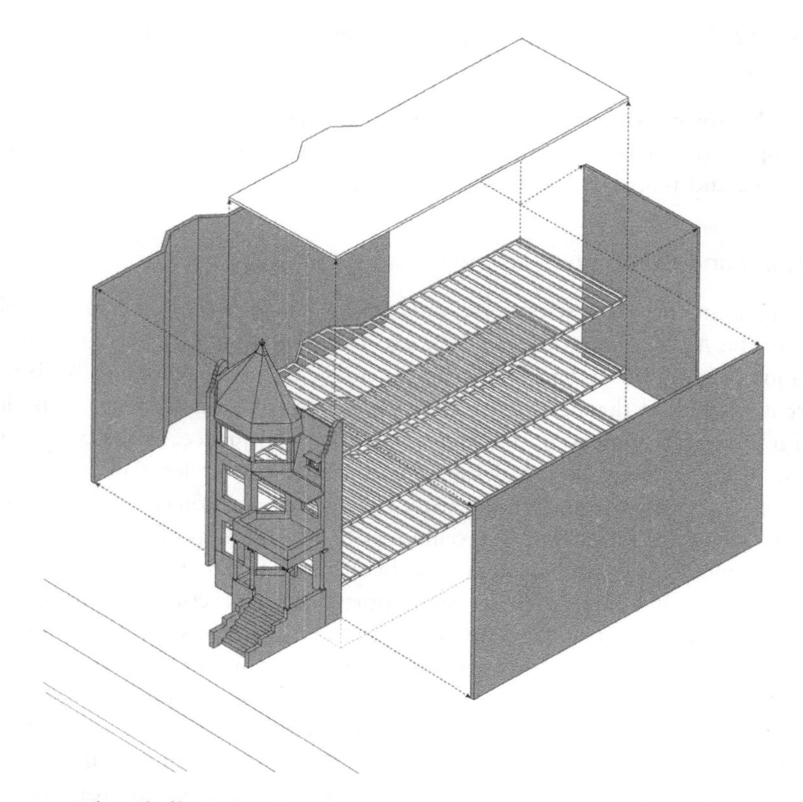

FIGURE 3.3 The *Phyllis Wheatley Home* historical facade and exploded view of the building enclosure.

FIGURE 3.4 Retrofit project of a single-family home in Chicago's South side.

FIGURE 3.5 The Growing Haven.

Integrated Design Approach

These courses aim for teams to develop a balanced and integrated design where architectural and engineering decisions mutually influence and benefit each other. In this process, initial design ideas and their impact on the building's energy performance are evaluated through energy models from the early design stages. The zero energy goal is met by integrating photovoltaic (PV) systems. Initial massing studies and the available roof area define the potential for on-site energy generation through a roof-mounted PV system. The target Energy Use Intensity (EUI) is calculated by dividing the annual estimated renewable energy generation by the total gross floor area. To achieve zero energy, the annual energy consumption must not exceed the annual energy generation. Thus, the design EUI must be equal to or less than the target EUI. The reduced energy demand is achieved by implementing energy conservation strategies, which constitute the first step in the design process. Although projects vary, they share common strategies, including optimized orientation and daylighting, natural ventilation, a high-performance envelope, efficient equipment, appliances and fixtures, and low embodied carbon materials.

InterTech,[19] a mixed-use student residence presented at the 2018 Race to Zero competition, exemplifies the integrated design approach. Two key features – the atrium and a split-level on the first floor – serve dual purposes, enhancing both architectural quality and building performance. Connecting different living areas, the atrium acts as a communal space for residents. Meanwhile it enhances natural ventilation. Likewise, the split-level on the first floor is motivated by the need to provide natural ventilation in the garage area beneath the southern portion of the building, thereby reducing the energy required to mechanically ventilate this space. This decision resulted in a spacious high-ceiling area on the building's north side, accommodating a café/grocery store. While energy-saving considerations initially drove this design choice, it enhanced the café's spatial quality by providing taller ceilings in the public area of the mixed-use program.

Additional strategies include an efficient space layout providing optimized access to daylighting, LED lighting, a well-insulated and airtight building envelope, efficient HVAC system and a ground source heat pump. Furthermore, the project's massing and roof tilt were designed to optimize on-site renewable energy generation through a roof-mounted PV system.

FIGURE 3.6 *InterTech*: exploded section view through atrium.

Assignments

The primary assignment for each team is to advance their project and present their progress at every class meeting. While part of the course involves lectures, most class time is devoted to project reviews. The class meets once a week for approximately 2.5 hours.

In addition, there are four assignments throughout the fall semester. The first involves conducting a literature review and analyzing a case study of a zero energy building. Working in small teams, students deliver written reports and five-minute presentations outlining the strategies the selected building design employs to achieve zero energy and their impact on the overall design. Typically, teams comprising mostly engineering students prioritize technical strategies in their presentations. On the other hand, architecture students may focus their presentations on the building's program and space layout. Understanding how each design decision is interconnected and affects other aspects of the project is an essential initial step for students to familiarize themselves with the project and competition.

The second assignment requires students to present project ideas for various sites and programs aligned with the Design Challenge divisions. Students explore multiple programs and potential locations. In class, each idea is collaboratively assessed, and the options are narrowed down to two or three projects, depending on the anticipated number of teams participating in the competition. Subsequently, the final competition teams are established. This assignment is crucial and may span more than one week. The number of anticipated teams depends on the class size, with team sizes typically ranging between 6 and 15 students. Furthermore, students are required to complete the DOE's Building Science Education Series and submit the first deliverable for the Design Challenge, which consists of a project summary.

In the spring semester, assignments primarily revolve around the competition's deliverables.

Additionally, during both semesters teams present their work twice to industry partners for midterm and final presentations and participate in Innovation Day.

Analytics

Students enter the class with varying backgrounds and levels of preparation. Architectural engineering students typically take courses in the Department of Civil, Architectural, and Environmental Engineering that establish a solid foundation for this class including Building Science, HVAC Systems Design, Energy Conservation in Buildings, Building Envelope Design, Architectural Design, Capstone Senior Design, Applied Building Energy Modeling, and Construction Methods and Cost Estimating. In these courses, students are exposed to various software used for:

1. analyzing the building enclosure performance;
2. daylight analysis;
3. energy analysis;
4. life cycle analysis, cost-benefit analysis and cost analysis; and
5. analyzing air flow patterns.

Students from the College of Architecture bring their expertise in architecture, interior design, and urban planning. They are proficient with drawing software and graphics, video-making, and rendering tools.

Throughout the semesters, experts are invited to familiarize students with some of these tools. Energy analysis is used from the initial stages to help inform and guide early design decisions. For example, it is conducted to compare various massing and orientation options to optimize energy performance and on-site energy generation potential. Additionally, energy analysis helps understand the trade-offs between different design decisions. For instance, increasing the window-to-wall area ratio may improve daylighting but negatively affect the building's overall energy consumption. Once the design is finalized in the second semester, teams develop more detailed energy models.

In addition to using energy analysis software, teams working on retrofits of existing structures have conducted field measurements, including surveys, developing digital twins, and taking thermal images of the building envelope to detect heat loss in the system.[20]

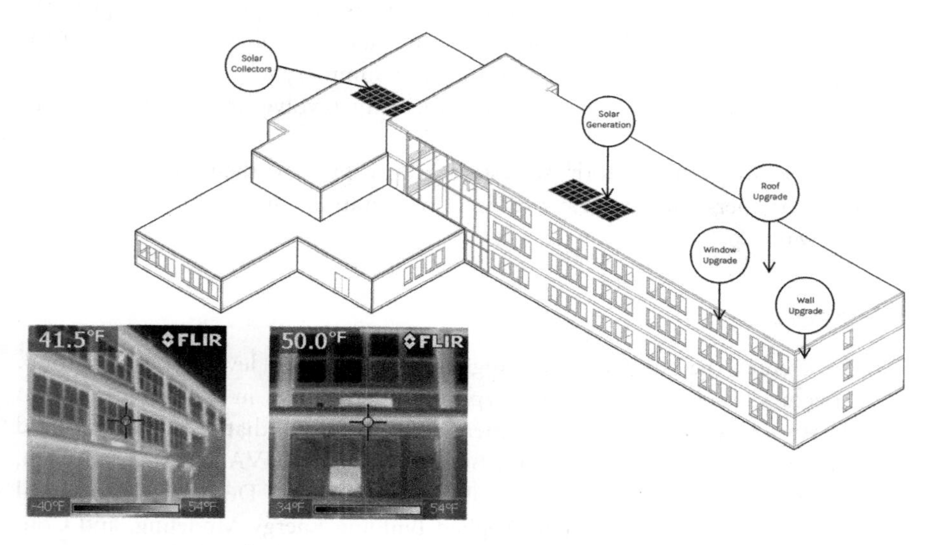

FIGURE 3.7 *ReDett Ed*: thermal images of the building envelope.

Required Readings

No textbook is required for this course. Required readings include:

Department of Energy, "A Common Definition for Zero Energy Buildings," prepared for the Department of Energy by The National Institute of Building Sciences, 2015, www.energy.gov/eere/buildings/articles/common-definition-zero-energy-buildings.

DOE Solar Decathlon Design Challenge Rules, www.solardecathlon.gov/sites/default/files/2025/assets/pdfs/design-challenge-rules-2025.pdf.

DOE's Solar Decathlon Building Science Education Series, www.solardecathlon.gov/education/building-science.

Solar Decathlon Design Challenge Past Project's Final Reports – 2024 Results, www.solardecathlon.gov/past/design/2024/results, earlier years available at same website.

Recommended Readings

Examples of recommended readings include:

ASHRAE: The Advanced Energy Design Guide – Achieving Zero Energy series.

The Built Environment Research Group (BERG), "Courses," https://built-envi.com/courses/.

Ching, Francis D. K. *Building Construction Illustrated.* 6th edition. Wiley, 2020.

Ching, Francis D. K., and Ian M. Shapiro. *Green Building Illustrated.* 1st edition. Wiley, 2014.

National Renewable Energy Laboratory (NREL) and Elevate Energy, "Achieving 50% Energy Savings in Chicago Homes: A Case Study for Advancing Equity and Climate Goals," 2020, www.nrel.gov/docs/fy22osti/83575.pdf.

Outreach Opportunities

Two articles focused on projects originating from Illinois Tech's Net-Zero Energy Building Design Courses:

1. Corradi Dell'Acqua, E., Marin, J., and Wright, E., "Integrated Architectural and Engineering Design Strategies for a Zero-Energy Building: Illinois Institute of Technology's Design Entry for the 2018 US Department of Energy Race to Zero (Solar Decathlon Design Challenge)," *Journal of Green Building* (2021) 16 (2): 251–270.
2. Kang, I., Maristany, Á. D., Corradi Dell'Acqua, E., Stephens, B., and Heidarinejad, M., "Assessing the Impacts of Natural Ventilation on Building Energy Use: A Case Study of a Mid-Rise Residential Building in Chicago, IL," Las Vegas, 2022.

The latter work has been honored with the ASHRAE Technical Paper award.

Moreover, projects developed within this course sequence were presented at the 2021 ASHRAE Winter Conference Seminar 7: "Building the Next Generation in Building Science: The Solar Decathlon Competition."

This work has also been presented at the ASES Solar 2024 conference, held at George Washington University from May 20 to 23, 2024.

Course Assessment

Integrating the Design Challenge into the curriculum has been successful, as reflected by the program's continuous growth. Over the years, Illinois Tech teams have achieved numerous awards in various divisions, including one honorable mention, one 3rd place, three 2nd places, and two 1st places.

Furthermore, the Master of Architectural Engineering and the Master in High-Performance Buildings were among the first cohort of post-secondary programs to earn the Department of Energy Zero Energy Design Designation (ZEDD) in 2022.

Looking Ahead

Enhancing the first part of the sequence by providing students with a more structured approach to learning the fundamental concepts of building science and their application in designing high-performance buildings is a key goal moving forward. Since design is an iterative process, it will help students develop an overview of the different steps of the process.

Furthermore, expanding collaboration across disciplines and forming partnerships with industry and community organizations remains a priority. Collaborating with industry professionals and mentors offers invaluable opportunities for students to gain real-world insights, access resources, and engage in meaningful projects and applications. These partnerships enrich the learning experience and better prepare students for their future careers.

FIGURE 3.8 *Liberty Arc.*[21]

Notes

1 "Interprofessional Projects (IPRO) Program | Illinois Institute of Technology," accessed May 2, 2024, www.iit.edu/academics/active-learning/ipro.
2 "Architectural Engineering (M.ENG.) | Illinois Institute of Technology," accessed May 2, 2024, www.iit.edu/academics/programs/architectural-engineering-meng.
3 "High Performance Buildings (M.HPB.) | Illinois Institute of Technology," accessed May 2, 2024, www.iit.edu/academics/programs/high-performance-buildings-mhpb.
4 "Solar Decathlon: About the Design Challenge," accessed May 2, 2024, www.solardecathlon.gov/2024/design-challenge.html.
5 "A Common Definition for Zero Energy Buildings," accessed May 2, 2024, www.energy.gov/eere/buildings/articles/common-definition-zero-energy-buildings.
6 "2016 Results," accessed May 2, 2024, www.energy.gov/eere/buildings/2016-results.
7 "Building Science Education Series," accessed May 2, 2024, www.solardecathlon.gov/building-science.html.
8 "Solar Decathlon: 2020 Design Challenge Participant Teams," accessed May 2, 2024, www.solardecathlon.gov/2020/design/challenge-results.html. *Vertical River* design team: Pablo Aranda Navarro, Fengyuan Jiang, Nupoor Nileshkumar, David Polzin, Rafael Ventura López and Sergio Vizcaino González.
9 Jonathan Ellison et al., "Coastal Refuge: A Flood and Hurricane Resistant Home," Solar Decathlon Design Challenge Project Narrative (Illinois Institute of Technology, 2024). *The Coastal Refuge* design team: Jonathan Ellison, Leonardo Godoy Bulnes, Babitha Gowda, Angel Morenilla Perez, Monique Perczynski, Beatriz Perez Olano, Julianna Roberts, Sarah Tessier, Saskshi Thakkar, Thomas Lozanovski and Aki Chang.
10 Industry partners who have supported the teams over the years include ASHRAE Illinois Chapter, Baumann Consulting, Callan Consulting Engineering, Civic Projects, Cushing Terrell, C|Z Design Strategies, dbHMS, Elevate Energy, HKS, JLL, Larson & Darby Group, Phius, SCB, Thornton Tomasetti and zpd+a Architecture.
11 "PRESERVATION CHICAGO – Love Your City Fiercely!," October 4, 2021, accessed May 2, 2024, www.preservationchicago.org/.
12 "Phyllis Wheatley Home – 2021 Most Endangered – PRESERVATION CHICAGO," March 4, 2021, accessed May 2, 2024, www.preservationchicago.org/phyllis-wheatley-home/.
13 Aamena Bakarmom et al., "The New Front Porch: Project Narrative," Solar Decathlon Design Challenge Project Narrative, 2022. *The New Front Porch* design team: Aamena Bakarmom, Nehemayah Green, Aaron Horta, Kexin Li, Erin Nelson, Sindhu Pandiaraj, Larry Tchogninou and Jonte' Williams.
14 "Sunshine Gospel Ministries – Home," accessed June 7, 2024, www.sunshinegospel.org/.
15 "Sunshine Gospel Ministries – Housing Equity Initiative," accessed May 2, 2024, www.sunshinegospel.org/hei.
16 Zeiad Sherif Mohammed Abdelkader Amin et al., "RetrofIIT," Solar Decathlon Design Challenge Project Narrative (Illinois Institute of Technology, 2023). *RetrofIIT* design team: Zeiad Sherif Mohammed Abdelkader Amin, Nora Awad, Emily Duong, Michael Graham, Ramisha Haque, Sydney Holles, Muntaha Imteyaz, John Kulka, Saumya Sukhthankar, Aung Mon, Patrick Dilger, Patrick Kubik, Rayyan Kiwan, Derek Pietrowski and Younghoon Jung.
17 Matteo Calafiura-Soleri et al., "The Growing Haven: A Mixed-Use Multifamily Building," Solar Decathlon Design Challenge Project Narrative (Illinois Institute of Technology, 2023). *The Growing Haven* design team: Matteo Calafiura-Soleri, Lucia Espinel Roig, Oana Giuglea, Nisha Kakde, Severin Kravchuk, Leighton Ledebuhr, Ewelina Robey, Riley Ross, Sarah Kay Stephens, Nhat Nguyen, Willa Vigneault and Lydia Davidson.
18 "Justice40 Initiative | Environmental Justice," The White House, accessed June 7, 2024, www.whitehouse.gov/environmentaljustice/justice40/.

19 Narjes Abbasabadi et al., "InterTech: Project Report Volume I and II," 2018. *InterTech* design team: Narjes Abbasabadi, Brett Horin, Vitoon Jittasirinuwat, Ajay Kotur, Jaime Marin, Esther Rodriguez and Eric Wright.
20 Zainab Amin et al., "ReDett Ed," Solar Decathlon Design Challenge Project Narrative (Illinois Institute of Technology, 2024). *ReDett* design team: Zainab Amin, Afrooz Asadollahzadeh, Wai Yan Aung, Pegah Ghofrani Esfahani, Kevin Garcia, So Jung Park, Rodrigo Salado, Stefannie Erin Tuason and Ashton Voorhees.
21 Robert Anderson et al., "Liberty Arc: Studio ZEB," Solar Decathlon Design Challenge Project Narrative (Illinois Institute of Technology, 2024). *Liberty Arc* design team: Robert Anderson, Paul Dhununjay Claude Chittoo, Ethan Gahan, Cameron Gluth, Nicolas LeDonne, Alexander Collins, BreAnne Long, Mariannys Lopez, Ethan Pulvermacher, Jayhawk Reese-Julien, Julian Sarria and Remi Edouard Jacky Thelier.

4

EXPANDING EQUITY AND EMPOWERING COMMUNITIES

Lessons from the Field

Julie Ju-Youn Kim

Course Overview

Course Name

Flourishing Communities Collaborative (FC2) Workshop

University / Location

Georgia Institute of Technology / Atlanta, GA, USA

Targeted Student Level

- Graduate
    - 2nd and 3rd year, Master of Architecture, 3.5-year program
    - 1st and 2nd year, Master of Architecture, 2-year program

Course Learning Objectives

Students who successfully complete the course will have shown their capacity to achieve all these objectives simultaneously. In this course, students will demonstrate:

- Ability to collaborate and share ideas and apply inventive approaches to problem finding and problem solving across scales and disciplines.
- Understanding of shared values of the discipline and profession, especially in relation to design, equity, diversity, inclusion and community engagement.

DOI: 10.4324/9781032692562-6

- Ability to address issues around climate change, energy burdens and community-centric education with actionable solutions.
- Ability to deploy appropriate representational modes and media in each of these aspects to advance the analysis and synthesis of design parameters and to communicate conceptual, technical and expressive intents.
- Effective communication with non-architect stakeholders via graphic and verbal means.

Course Learning Outcomes

Successful completion of this course will enable students to:

- Effectively and successfully engage in collaborative and participatory community engagement processes.
- Advance design and applied research towards actionable strategies centered on climate change, energy burdens, and social and environmental concerns.
- Integrate the diverse needs, social and spatial patterns that characterize different cultures and individuals.
- Expand the responsibility of the architect to ensure equity of access to sites, buildings and structures.

Course Methodology

Philosophy

To offer broader context to my philosophy, I must start by sharing my path that supports my current work in the space of equity and inclusion through design. Through this chapter, I will underpin how important and relevant this work is, particularly in the shared space of education and practice. In my mind, it is important to talk about both *why* and *how* we do it.

I have always operated with a deep understanding and awareness of being "different." I am a woman architect. I am a Korean-American architect. In some ways, I have always operated from the edges. However, rather than start from a place of "less-than," I switch the narrative to start from a place of empowerment. I establish strategic partnerships and pioneer educational opportunities that situate myself, my students – and architecture – in the middle. Let me offer this redefinition of myself. I am an architect-educator deeply invested in expanding equity through design. This is part of the "why," articulated as a series of strategic aims:

- Advance entrepreneurial research and practice via strategic partnerships.
- Support innovative pedagogical frameworks.
- Leverage our capacities as design thinkers across disciplines.
- Expand the role of the architect as an entrepreneur, researcher, strategist, practitioner, activist and inventor.

The combination of each of these strategic aims offer context to my research and design lab, Flourishing Communities Collaborative (FC2). By offering experiential learning opportunities through coursework with the support of dedicated and committed community partners, our project offers students real-world exposure to the practice of Public Interest Technology. We have built and sustained partnerships with nonprofit and affinity group partners, sharpening the real-world application of Public Interest Technology to center community voice in pursuit of solutions for pressing problems, particularly challenges experienced by marginalized communities least well served by existing systems and policies.

Through design studios and workshops, I am deeply invested in demonstrating the value of a quantitative, analytical and computational approach to the social and cultural functions of the built environment.

Definitions

Carbon neutrality (in the space of social responsibility) – Our work considers goals towards carbon neutrality can be met by centering community voice in pursuit of solutions for pressing problems, particularly challenges experienced by marginalized communities least well served by existing systems and policies.

Design (verb) – It is a projective act, a method.

Flourishing community (systems) – For a community to flourish and be sustainable, they must have access to five fundamental things: housing, education, jobs, healthcare and transportation.

Public interest technology (systems) – "At its heart, public interest technology means putting people at the center of the policymaking process – not just by designing programs with constituents' needs in mind, but by engaging directly with constituents throughout the policy design and implementation process."[1]

Implementation

The Georgia Tech 2020–2030 Strategic Plan includes focus areas on *expanding access, amplifying impact, leading by example.* Academic programs hold the unique opportunity to serve as consultants and advocates for communities that may not have the resources necessary to hire design services. We can, likewise, highlight issues around design access and equity. Academic programs do not have any skin in the game, so to speak. Thus, with complete transparency, we provide crucial historical and regulatory information, presenting communities with all the facts for making informed and sustainable decisions instead of being swayed by influential parties who may have political or development interests.

FIGURE 4.1 Graduate students sharing architectural proposals to young community stakeholders.

Source: Photograph by author

In FC2, we naturally align with the Georgia Tech Strategic Plan focus areas in these ways:

- We *amplify the impact* of our future leaders in practice by preparing students to apply inventive approaches to problem finding and problem solving across scales and disciplines.
- We *connect* students to practitioners who model best practices in communication, collaboration and design thinking.
- We *lead by example* as we engage in teaching and serving by working with communities.
- We *empower communities* by expanding access and using technology as a means to bring joy and hope through the optimistic acts of making and design.

FC2 connects the academy and practice – it is a platform for engagement where teaching and learning are seen as reciprocal activities. I offer a curricular working model around social carbon neutrality. In other words, through deliberate collaborations, our applied research and design lab explores and implements initiatives centered on increasing access and promoting equity.

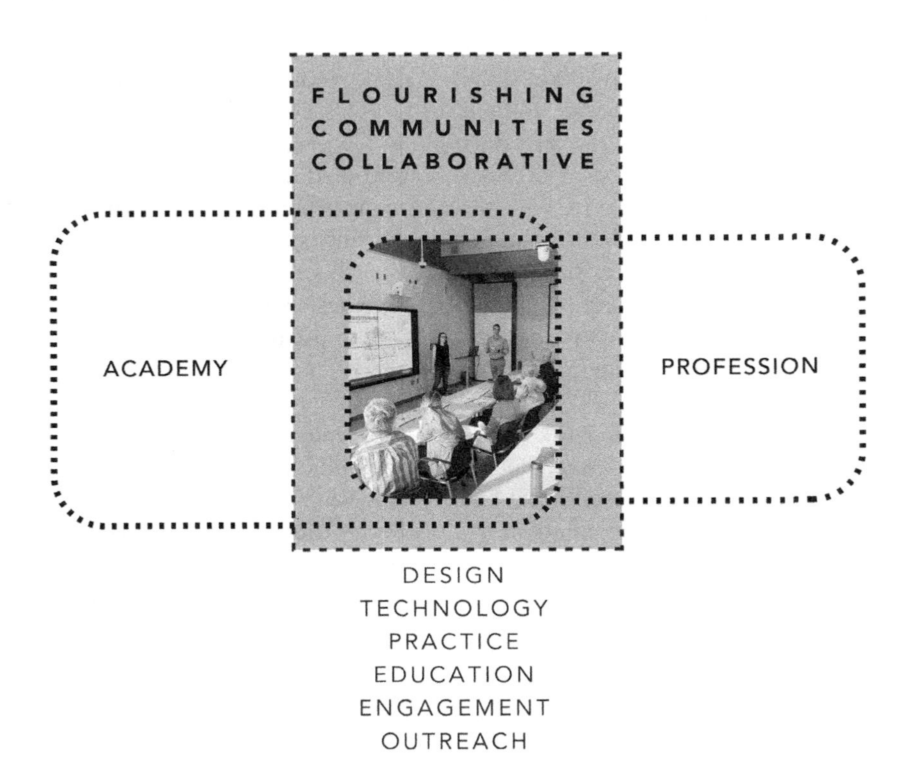

FIGURE 4.2 Diagram illustrating FC2 connections across academy and practice.
Source: Diagram by author

I need to note here that I am not an expert. I am learning as I go, from both our successes and our failures. In fact, it is our failures that can have the most impact. In other words, when we hit a bump or a wall, this encourages more creative thinking to re-direct towards the final desired outcomes. Sometimes these outcomes change, too ... and that's okay. There have been important lessons learned as we navigate *how* we do our work.

- Working with community is hard – it requires the patient and determined work of building and maintaining trust. It is hard to cultivate – and so easy to lose.
- Trust takes time and patience, and we must be respectful and supportive of the process. Community engagement cannot be a new form of colonialization, and one-sided extraction and advertisement. As my colleague Danielle Willkens has sharply pointed out, the community itself needs to lead with their goals and ambitions, and programs can present possible options or unexplored alternatives. Ultimately, it must be an integrative process from the inception of the project or intervention.
- Effective community relationships – this is different from "engagement" – are nurtured by listening. I like to tell my students that our work includes creative listening which leads to creative thinking and making.
- Communities – particularly disadvantaged communities – are not looking to be "fixed" – they want to be heard, to be seen, to be rendered visible!
- Site-based learning and investigation is the very basis of all our work. Research sits at the core of work. Instead of directly jumping into design activities, we must start with the questions around how and why certain conditions exist and what can be done to shift systemic issues.
- Patience. We must resist the impulse of quick solutions. These tend to not have the lasting power required to ensure consistent change.
- Finally, technology is not the cure. We aim to use technology as a means to empower the community.

The true vision for our work is grounded in equity. Housed in the School of Architecture, College of Design, Flourishing Communities Collaborative (FC2) is a fully integrated teaching model that builds relationships between students, faculty, practitioners and the community in applied design thinking and problem solving for communities. Across scales, our efforts foster and support citizen-centric social impact and equity. Our fundamental aims are to empower a community with tools directed at investment of local expertise, amplify impact for underserved community members, and offer a framework for smart, sustainable development. We are committed to advancing public interest to support, specifically, members of society who have been historically excluded from such benefits. We connect technology, design, design-thinking and education offering a platform to study the capacity technology offers as a tool to address socio-technical, economic and cultural challenges, specifically for disadvantaged communities.

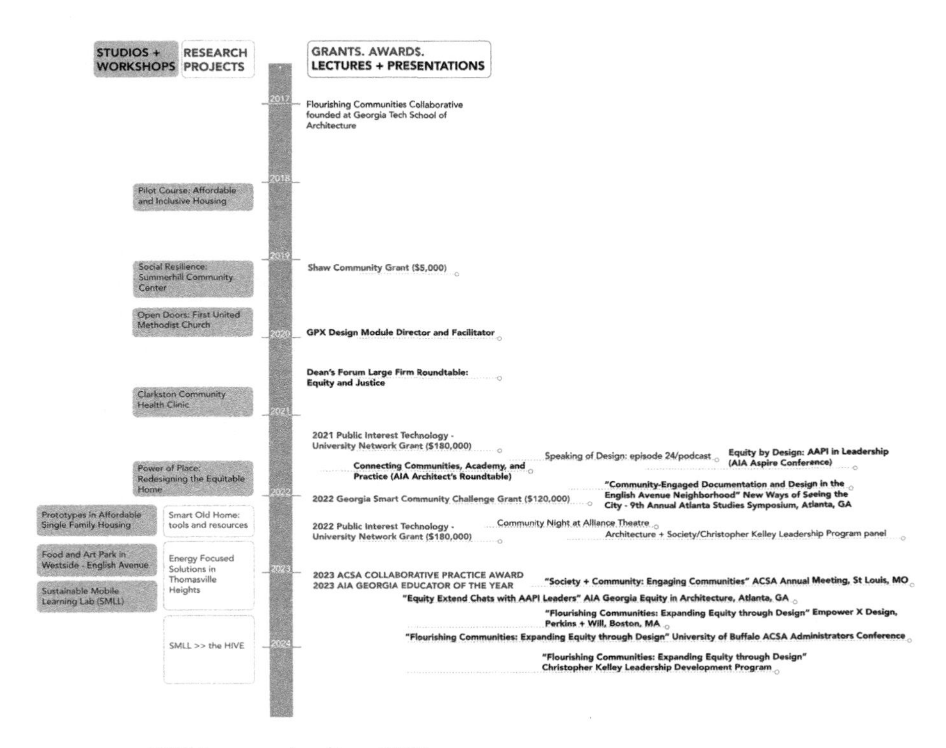

FIGURE 4.3 FC2 Impact timeline, 2017–present.
Source: Diagram by author

Assignments

In alignment with the five course learning objectives, FC2 courses intentionally frame the pedagogy across each of the learning objectives. Following interactive workshops with community stakeholders, I flip the script with the students. Rather than outlining the next steps for them, I ask them to tell me what they heard. As we share our collective reflections, the next assignment takes shape. Via active intentional engagement, students develop their abilities to collaborate, share ideas and apply inventive approaches to problem finding and problem solving across scales. This type of teaching methodology fosters students' development of invaluable skills in creative listening – this is something they can take with them in future academic and, certainly, professional pursuits. Listening is just one part of the puzzle – the other part is crafting outputs that effectively communicate conceptual, technical and expressive intents to a diverse audience. What is of tremendous value is that students not only gain firsthand experience in understanding how to clearly communicate across disciplines but, equally importantly, they develop deep understanding of shared values around design, equity and community engagement.

Our initiatives wrestle with issues around climate change, energy burdens and community-centric education. Our recent work over the past two and half years centers in a specific area[2] of Atlanta on projects that are all inter-related – this includes a prototype for a single family affordable home that integrates building performance to alleviate passing energy burdens onto the homeowner; a digital resource with co-designed tools for current homeowners aimed at expanding equity by expanding knowledge; a sustainable mobile lab that is, in itself, a teaching tool about construction and sustainability – with need-based programming such as pop-up vaccination clinics, resources for free mammograms, resources for voter education and rights, fresh food, DIY workshops, and so on. This is all the important work of empowering communities through design. FC2 extends the projective nature of the act of design by sharing tools with the community and they are, in turn, equipped to apply these to the specific issues that are important to them.

Analytics

I will share three projects to illustrate our approach to analytics and software in design development.

Project One: Smart Old Home

Atlanta has the 4th highest energy burden in the country, where the residents pay more than 10% of their family income on energy bills. Our multi-institutional team assembled and designed an online resource for current and future homeowners in disadvantaged communities.

For our 2021 PIT-UN project, Smart Old Home (https://fc2.design.gatech.edu/soh/), our team developed a robust website that houses a set of tools and

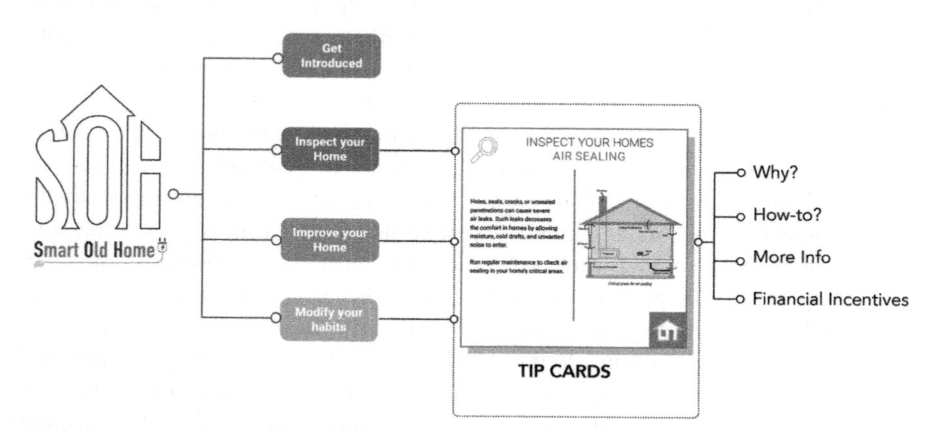

FIGURE 4.4 Smart Old Home navigation tree diagram.
Source: Diagram by FC2, Georgia Tech School of Architecture

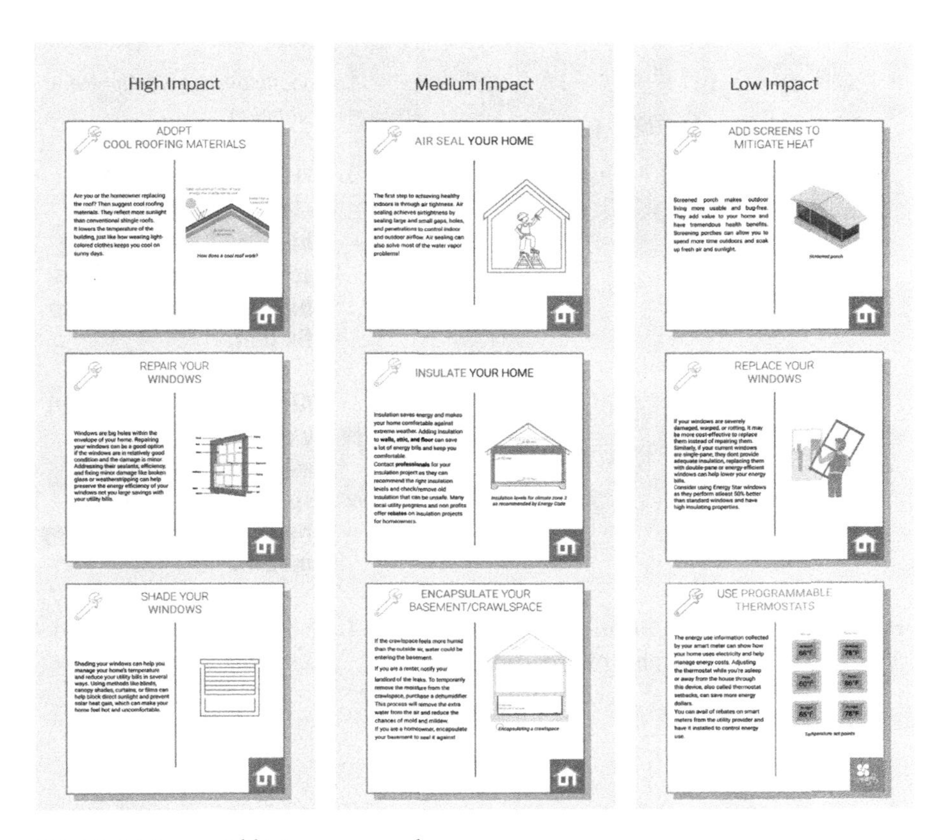

FIGURE 4.5 Smart Old Home tip cards.
Source: Illustration by FC2, Georgia Tech School of Architecture

methods to address problems related to energy burdens carried disproportionately by disadvantaged communities. We learned that there are many current homeowners who lack the basic knowledge of how to manage their energy consumption from simple behavioral modification to more ambitious maintenance updates across scales. We organized sets of tools across three possible actions: inspect, improve, modify – and developed a clear and accessible structure that homeowners may easily navigate. This set of tools we designed equips the homeowner with knowledge (education) and resources that will help build economic and environmental sustainability.

Project Two: The HIVE

Sustainability is a core element of our projects. *The HIVE (Housing Innovation: a Vehicle for Education)* (https://fc2.design.gatech.edu/sustainable-mobile-learning-lab-for-westside-neighborhood-english-avenue/) seeks to celebrate aspects of the diverse neighborhood identity by offering a platform for shared experiences, open

Joann desires a better living situation.

"I am just getting by - I cannot save money because I am spending so much on my utilities, but I don't know how to fix that.

Also, if my house layout were more flexible, I might be able to rent space and this would help me a little bit every month."

FIGURE 4.6 Photos from the neighborhood and notes from a community stakeholder.
Source: Photos and graphic by author

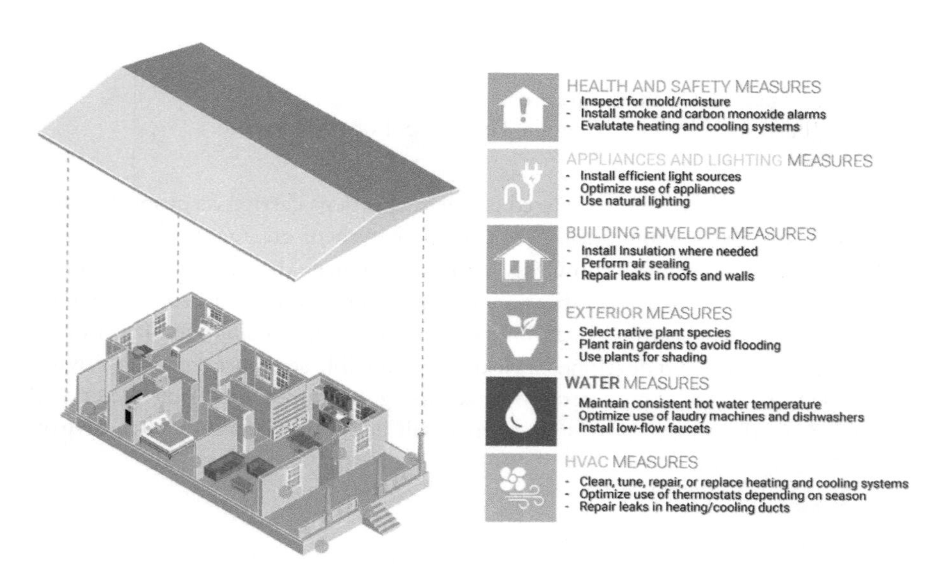

FIGURE 4.7 Axonometric diagram illustrating key areas for targeting economic and environmental sustainability.
Source: Diagram by FC2

FIGURE 4.8 Collage study illustrating opportunities in the Westside Neighborhood – English Avenue.

Source: Image created by Monica Rizk and Rand Zalzala, Georgia Tech School of Architecture

engagement, exchange and education for an under-resourced community. A collaboration that includes students, faculty, community stakeholders, an architect/practitioner, a structural engineer, a community non-profit and a community builder/developer, the HIVE focuses on social infrastructure needs in the community and offers a mobile learning space for the community, ensuring equitable access to the broad set of constituents in the neighborhood.

In a beta-test build, the HIVE will operate as a demonstration model for what could be available for communities as a living learning laboratory, introducing communities to topics around sustainability, resilience and circular economies. Through this project's design and execution, we introduce architectural practice to younger residents of the communities around Georgia Tech, enriching the educational pipeline for underrepresented students. Additionally, the HIVE will offer a pipeline for community residents of all ages to learn about history, architecture, technology, sustainability, construction and planning.

The HIVE serves as a beacon of eco-consciousness, showcasing the potential of responsible construction practices. By integrating energy-efficient features such as optimal shading and optimized air flow into its design, the project sets an example for environmentally friendly development. Comfort and flexibility are carefully woven into the very fabric of the HIVE. The project incorporates shading and wind studies to optimize outdoor shaded areas and create a

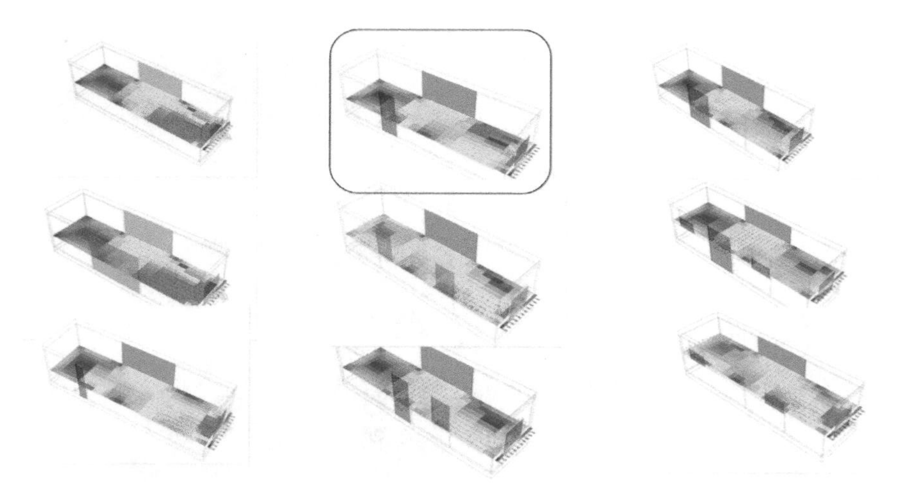

FIGURE 4.9 CFD analysis to determine optimal locations for openings.
Source: Courtesy of FC2, Georgia Tech School of Architecture

FIGURE 4.10 Sharing the development via physical models and VR of the HIVE with community stakeholders, Summer 2023.
Source: Photo by FC2, Georgia Tech School of Architecture

comfortable interior atmosphere. A lightweight diagrid canopy, made of steel and polycarbonate cladding, provides shade and shelter, symbolizing the project's resilience and unity. Resting on a trailer bed, the structure can easily move to different locations within the neighborhood, allowing it to adapt to changing community needs and preferences.

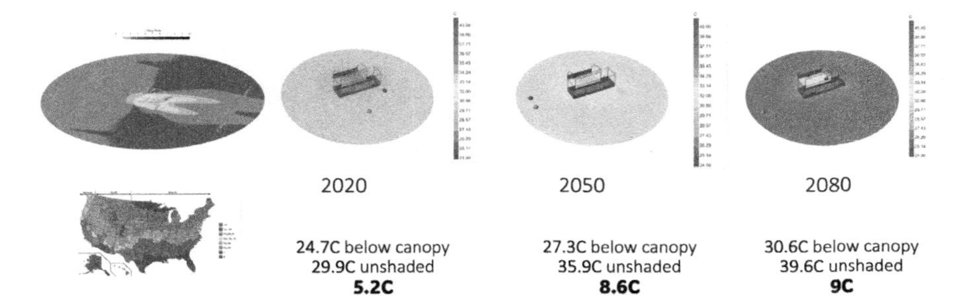

FIGURE 4.11 Data analysis illustrating the impact of shaded spaces relative to increasing temperatures due to climate change.
Source: Diagram by Max Doersam, Georgia Tech School of Architecture

FIGURE 4.12 Model study of the HIVE, summer 2023.
Source: Photo by author

Project Three: Energy-Efficiency Retrofits in Thomasville Heights

Another one of our most recent projects employs innovative diagnostic techniques that hold promise for achieving significant cost savings compared to traditional building energy auditing practices. By analyzing the building envelope exterior, using drones equipped with remote sensing instruments, this method is

FIGURE 4.13 Map diagram illustrating location of the Thomasville Heights neighborhood relative to downtown Atlanta.
Source: Diagram by author

also minimally invasive, avoiding much of the labor-intensive inspection of the home interior that is typically required during the energy audit phase. In this sense, it holds promise for overcoming homeowner hesitancy during the marketing and recruitment phase of weatherization programs. This project is especially timely in disadvantaged communities, where ongoing challenges such as acute unemployment and poverty will soon be compounded by the closure of a long-neglected subsidized housing project and the significant population loss that it entails.

After quantifying the difference between pre- and post-retrofit data collected for each data type, we saw overall improvement in every category on average – the greatest improvement seen in occupant comfort at 92% and least improvement in energy use at 9%.

Our pilot phase of the project prioritizes cost efficiency, striving to leverage grant dollars to achieve the greatest possible impact. Our aim is to broaden awareness so that future iterations of the program would be intentionally designed, drawing on local partners with expertise in equitable development, to prioritize minority business enterprises (MBEs) and other disadvantaged business enterprises (DBEs) in the contracting phase, particularly those local to the service area. Key takeaways from this pilot effort include the following:

- Low-budget retrofits and weatherization makes a tangible positive impact on vulnerable communities and is the first step towards preserving legacy homeownership.
- A single, sustainable and reliable weatherization program is necessary for community members to trust and recommend to their neighbors. It must also continue beyond one year.
- Weatherization has the potential to contribute to long-term benefits such as health, comfort and building envelope integrity.

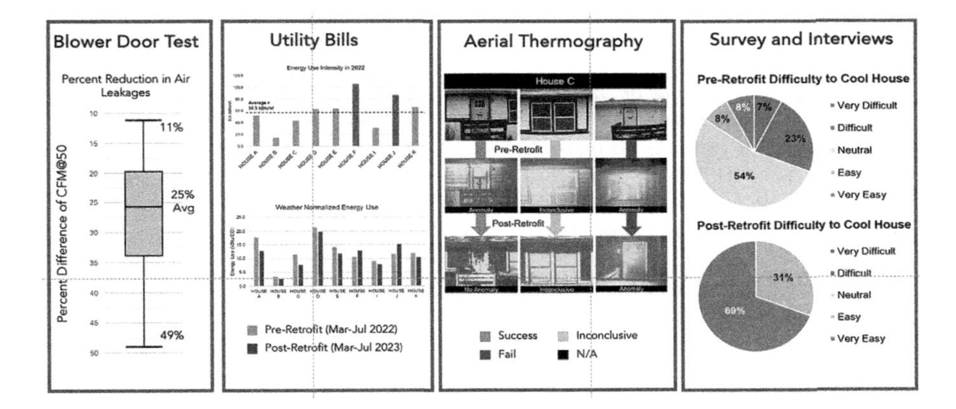

FIGURE 4.14 Pre- and post-retrofit data analysis. This offers a snapshot of the data we collected pre- and post-retrofit for comparison so we could determine the impact of the weatherization. This example of aerial thermography for one of the houses takes the regular image, compares it with the pre-retrofit image to determine any anomalies, and then compares it with the post-retrofit image to see if the anomaly has been mitigated or not, or if it is inconclusive.

Source: Diagram by FC2, Georgia Tech School of Architecture

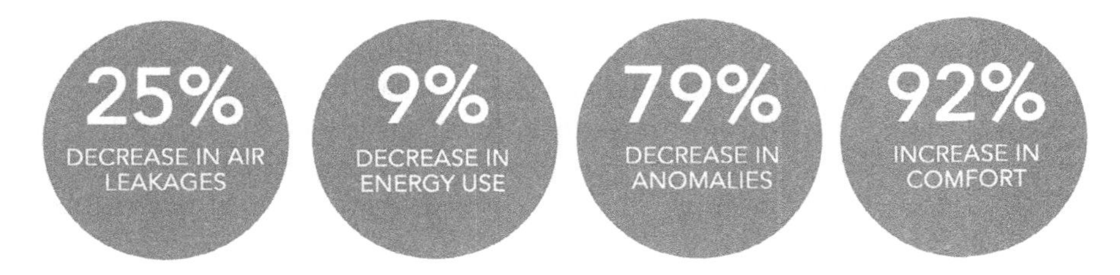

FIGURE 4.15 Pre- and post-retrofit data analysis. Low-budget retrofits and weatherization clearly have a tangible positive impact on vulnerable communities, and is the first step towards preserving legacy homeownership.

Source: Graphic by FC2, Georgia Tech School of Architecture

Required Readings

Armborst, T., D'Orca, D., & Theodore, G. (2021) *The Arsenal of Exclusion & Inclusion.*
Haselmeyer, S. (2023) *The Slow Lane: Why Quick Fixes Fail and How to Achieve Real Change.*
Hood, W., & Tada, G. (2020) *Black Landscapes Matter.*

Recommended Readings

This is a list of teaching references that I share with my students to build foundational thinking – it is an evolving resource list.

Bergdoll, B., & Christensen, P. (2008) *Home Delivery: Fabricating the Modern Dwelling.*
Bull M., & Gross, A. (2018) *Housing in America: An Introduction.*
Burdett, R. (2007) *Living in the Endless City* [The Urban Age Project by the London School of Economics and Deutsche Bank's Alfred Herrhausen Society].
Calvino, I. (1972, trans. 1974) *Invisible Cities.*
Habraken, N. J. (1998) *The Structure of the Ordinary: Form and Control in the Built Environment.*
Harris, D. (2013) *Little White Houses: How the Postwar Home Constructed Race in America.*
Kieran, S., & Timberlake, J. (2004) *Refabricating Architecture.*
Salazar, J., & Gausa, M. (1999) *Single-Family Housing: The Private Domain.*
Wachsmann, K. (1995) *Building the Wooden House: Technique and Design.*

Outreach Opportunities

Challenges should be embraced early and often. The world is full of problems that require a broad cross-section of disciplines to develop and propose solutions. There will be teams that include biologists, engineers, biophilic engineers, planners, mobility consultants, computer scientists, anthropologists, sociologists and architects. *How do we prepare our students to effectively participate on such teams? Where is our common ground?* "Consequently, we as designers, or as designers of design processes, have had to be explicit as never before about what is involved in creating a design and what takes place while the creation is going on ..."[3] Ultimately, we (the architects) can and should define, describe, even design our roles. As architects, we can be the connective tissues between coarse and fine grain thinking, offering value to multi-disciplinary problem solving leading to possible new directions. It is our elasticity that makes us uniquely equipped to address complex problems and to potentially expand the scope, and even the definition, of design outside the traditional architectural design studio.

Outreach is broad and diverse, including community engagement sessions; student leadership presenting to city and other government officials; and lectures and workshops to allied disciplines. Since 2021, my students and I have presented twenty times nationally in invited lectures, panel discussions and/or workshops.

FIGURE 4.16 Images from outreach sessions with community stakeholders, 2017–2024.
Source: Photos courtesy of author

Course Assessment

The coursework offered under the umbrella of the Flourishing Communities Collaborative sits within two interrelated and equally important frames: (i) knowledge and innovation, and (ii) leadership, collaboration and lifelong learning.

FC2 prepares students to engage and participate in architectural research to test and evaluate innovations in the field. Our work ensures students understand approaches to leadership in multidisciplinary teams, diverse stakeholder constituents, and dynamic physical and social contexts, and learn how to apply effective collaboration skills to solve complex problems. Finally, and most importantly, our work furthers and deepens students' understanding of diverse cultural and social contexts and helps them translate that understanding into built environments that equitably support and include people of different backgrounds, resources and abilities.

I advance the shared core value of knowledge and innovation[4] by offering curricular opportunities to advance the discourse around the future of Architecture as a creative, technical and intellectual interdisciplinary pursuit that requires integrated knowledge from various disciplines organized by structured collaborative teamwork. Another shared core value around leadership and collaboration[5] demands lifelong learning, which is a shared responsibility between academic and practice settings. These two frames situate multiple models of engagement centered on promoting equity and increasing access. I position curricular opportunities to advance the expanded role of the architect as practitioner, strategist, activist and entrepreneur.

As an educator, I connect practice and education, preparing future practitioners and scholars for the challenges of a constantly evolving global marketplace. Founded in 2017, FC2 sharpens the real-world application of public interest technology, which is defined as a set of practices to design, deploy and govern technology to advance public interest. We have demonstrated consistent

Can we imagine a role for the architect where the act of design is an entrepreneurial and innovative endeavor - for those who are underserved and lack access to resources?

FIGURE 4.17 Georgia Tech graduate students at a community engagement event with stakeholders.
Source: Photo courtesy of author

effort in working towards meaningful change. FC2 is oriented towards interdisciplinary focus by bringing together collaborations between researchers and scholars at Georgia Tech as well as from other universities, government agencies, community partners, professional practice, faculty and students.

Our work has received recognition from external peer organizations. Since 2016, I have secured $415,000 in grant funding; recognized with the 2023 AIA Georgia Educator of the Year Award (specifically noting my contributions to education via FC2 engagement); and the 2023 ACSA Collaborative Practice Award for FC2's projects, entitled *The Power of Place and Social Production.*

Looking Ahead

I always say that our work is just getting started – I envision continuing to expand the footprint of FC2 with community-engaged initiatives aimed at building knowledge and broadening access. Fundamentally, the projects forming the backbone of FC2's curricular framework include the design and research for a mobile learning lab; co-designed tools offering a resource for homeowners

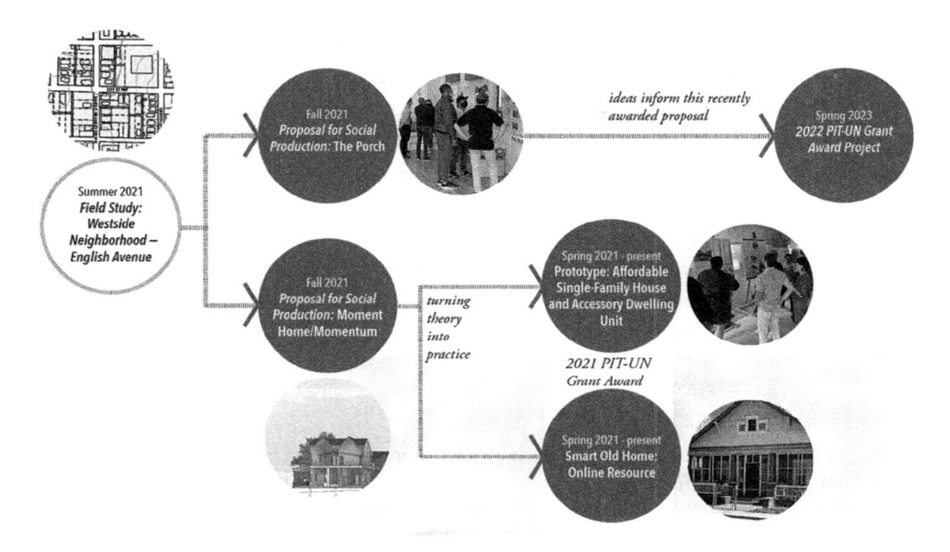

FIGURE 4.18 Graphic timeline of funded projects in Westside Neighborhood – English Avenue, 2021–2024.
Source: Diagram by author

to manage energy burdens; and a pilot weatherization program aimed at improving energy efficiency in older homes. Each of these projects expand access to knowledge, thus empowering the community with skills to support a sustainable future. The output is scalable and replicable, grounded by the idea that the tools may apply to other similar communities.

We work directly with community stakeholders through direct engagement of the processes of investment in local expertise, design and construction. Through each project's design and execution, we introduce architectural practice to younger residents, enriching the educational pipeline for underrepresented students. Community stakeholders support and contribute to the programming geared towards strengthening social and economic capacity for residents. FC2 curricular opportunities expand a pipeline for community residents of all ages to learn about history, architecture, technology, sustainability, construction and planning.

Acknowledgements

I must recognize and thank all of the individuals and organizations who have made our work in the Flourishing Communities Collaborative possible. I am deeply grateful to the following organizations: Georgia Tech School of Architecture and College of Design; Public Interest Technology-University Network; Partnership for Inclusive Innovation; Westside Neighborhood – English Avenue; OaksATL; Thomasville Heights/Norwood Manor communities; Focused Community Strategies; and the Beloved Community. There are many individuals I need to recognize as well. I could not have even begun this work without champions from the community – Winston Taylor, Mother Mamie Moore, James Tomlin – thank you for your commitment and engagement. This is a group effort, and I am so grateful for my incredible colleagues in the School of Architecture at Georgia Tech – Associate Professor Danielle S. Willkens and Associate Professor Tarek Rakha. And, most importantly, all of the students, starting with the first pilot course in 2017, who have joined me on this journey to leverage design as a tool for social justice and equity, demonstrating the value of a computational approach to social, environmental and cultural issues in our built environment.

Notes

1 New America, "Public Interest Technology: About," www.newamerica.org/pit/about (accessed March 25, 2024).
2 Today, approximately 44% of homes in the surrounding English Avenue Neighborhood are vacant and two-thirds of the residents live below the federal poverty line. Urgent neighborhood concerns include the existing service desert, hydrology challenges exacerbated by climate change and mounting external development pressures. However, the neighborhood is poised for positive change through the Westside Land Use Framework Plan (2017), already approved by the community and city. In a historically vibrant, but rapidly shrinking neighborhood, our work will set positive steps forwards to bolster retention and help in stabilizing the community.
3 Herbert A. Simon, *The Sciences of the Artificial*, third edition (Cambridge, MA: MIT Press, 1998).
4 Knowledge and Innovation. A core value shared by the National Accrediting Architectural Board (NAAB), The American Institute of Architects (AIA), the Association of Collegiate Schools of Architecture (ACSA), and the National Council of Architectural Registration Board (NCARB) is that the new knowledge advances architecture as a cultural force, drives innovation, and prompts the continuous improvement of the discipline.
5 Leadership, Collaboration, and Lifelong Learning. Another core value shared by the NAAB, AIA, ACSA, and NCARB is that the practice of architecture demands lifelong learning, which is a shared responsibility between academic and practice settings.

5

BUILDING DECARBONIZATION

Theory to Practice

Nea Maloo

Course Overview

Course Name

Building Decarbonization: Theory to Practice

University / Location

Howard University / Washington, DC, USA

Targeted Student Level

- Graduate
  - 4th and 5th year, Master of Architecture, 5-year program

Course Learning Objectives

In this course, students will:

- Learn basic building decarbonization (BD) concepts and discuss their impact on the design process, as well as energy terms like EUI, Btu/sf/yr., kWh.
- Gain awareness of building systems such as low energy, all-electric, building photovoltaics and grid interactivity in the context of BD.
- Understand and apply embodied and operating carbon calculation using software in the designs.
- Learn about retrofitting existing buildings with appropriate design and materials.

DOI: 10.4324/9781032692562-7

- Select and create appropriate architectural designs for existing buildings.
- Select materials with low carbon impact and considerations to create BD.
- Learn different methodologies for carbon offset and renewable energy applications.

Course Learning Outcomes

Successful completion of this course will enable students to:

- Identify the steps required to decarbonize buildings and distinguish the architectural features needed for low energy buildings.
- Analyze the different lighting and daylighting strategies and compare the different mechanical systems suitable for the project to be electric.
- Discuss how building materials have carbon impact and develop a decarbonized building design.

Course Methodology

Philosophy

This is a timely course about an urgent topic. The course philosophy mirrors the experience of working as an architect on the built environment with students who have global perspectives of decarbonization and awareness of energy justice. This course empowers these students to participate in real-world solutions as soon as they graduate.

The course material was developed in collaboration with Stanford University's Building Decarbonization Learning Accelerator (BDLA), which, among other things, provides free resources on how to teach decarbonization. As a result, the outcomes of this course are about goal achievement within the syllabus and, importantly, real-world skill-building for students whose graduations at the fourth- and fifth-year of an M.Arch program are right around the corner. Successful completion of this course enables students to identify the steps required to decarbonize buildings and distinguish the architectural features needed for low energy buildings. Successful completion of this course also enables students to analyze the different lighting and daylighting strategies and compare the different mechanical systems suitable for the project to be electric. Finally, successful completion of this course enables students to discuss why building materials have carbon impact and to develop a decarbonized building design.

The planning and course development for "Building Decarbonization: Theory to Practice" was a response to the social movement of Black Lives Matter for the empowerment of the next generation of Howard University students to combat climate change. The course was planned and developed in collaboration with Stanford University professors and Howard University professors as a new course in the architectural curriculum at Howard University, with inter-university collaboration as an example towards more innovative teaching and environmental justice. The design class was also created as a summer bridge course to upskill the graduating class for learning the theory and practice of decarbonization in 2021. The course has also been successfully implemented with a growing student cohort in the following three summer sessions.

Definitions

Carbon neutrality – The path to carbon neutrality is computing carbon, applying carbon reduction measures and carbon offsetting measures such as sequestration and carbon sinks. The net objective is to emit no more carbon into the atmosphere than is absorbed.

Coefficient of performance (COP) – A measure of an equipment's efficiency. For example, a COP of 3 means that for every 1 unit of electricity in, 3 units of energy is output (or, 300% efficient).

Decarbonization – The process of reducing carbon emissions in the atmosphere by human activity. Since buildings are responsible for approximately 40% of global emissions, we as designers and educators can have a larger impact in reducing the emissions.

Energy use intensity (EUI) – The total annual energy use for a building in kBtu/SF/year.

Life cycle assessment (LCA) – A method for analyzing the environmental impact of a material or product over its entire life from manufacturing to recycling or disposal.

Life cycle cost assessment (LCCA) – A method for comparing the costs vs. savings of a material or product over its entire life from manufacturing to recycling or disposal.

Implementation

General

To deal with the complexity of decarbonization, this course breaks it down into a simple methodological framework and creates assignments that students can complete with simple software in a sequential, linear fashion. The goal is to help them build confidence in acquiring and "speaking" the technical language of both design and sustainability, as well as to give them a way to apply that language to the real-world problem of the renovation of the headquarters of the American Institute of Architects (AIA) at 1735 New York Avenue, NW, in Washington, DC, a concrete, steel and glass building designed by The Architects Collaborative more than 50 years ago and in need of upgrades. Its importance is both as a symbol of the Country's premier professional architecture association and as a workplace for more than 200 people.

FIGURE 5.1 AIA Headquarters in Washington, DC.
Source: Nia Baptise

The students learn about small sections of the building and apply their skills to actual design problems related to those sections. The students also engage in discussions of what is equity and justice as they iterate different solutions to renovating this structure. The course has several guest speakers, including the architect for the AIA HQ building renovation and other industry experts. The use of current case studies discussed in the course are written by professionals working on the AIA HQ building currently and who offer students the background they need to evaluate challenging concepts like building lifespan, the influence of codes on design and building practices, and what it means to not just think about decarbonization, but to pursue strategies for net-zero carbon neutrality. The students then use this research in their final year thesis projects as a way of expanding their knowledge base. Through the process, students cover several major sections including an overview, architectural considerations, lighting and daylighting, embodied carbon, building systems and electrifications. solar PVs, and calculations. For the final deliverable, the students work collaboratively to connect these pieces together into a single cohesive decarbonization strategy for the AIA's Brutalist building.

Renovating Brutalist architecture to meet modern energy efficiency standards presents a unique set of challenges due to the design principles inherent in these structures. Brutalist buildings, characterized by their imposing concrete forms and stark geometries, are celebrated for their bold aesthetics and robust construction. Yet, these buildings often lag behind in terms of energy performance compared to contemporary standards. Concrete, the primary material in Brutalist buildings, possesses high thermal mass but lacks effective insulation. This results in significant heat loss during winters and heat gain in summers. Retrofitting insulation without compromising the building's aesthetic integrity is a challenge. External insulation systems or interior insulation systems must be carefully integrated, considering the building's unique geometry and surface textures.

Brutalist buildings sometimes feature small windows or narrow slits, reflecting the architectural preference for fortress-like solidity. This is not so at the AIA HQ building, which includes large amounts of fenestration along the southern and southwestern fronts, often creating a greenhouse effect for those sitting next to them, juxtaposed with the complete lack of daylight on the northern and northeastern fronts. Enhancing daylighting while minimizing solar heat gain requires thoughtful window replacement or enlargement. However, altering window sizes can disrupt the building's visual coherence and raise structural concerns. Advanced glazing technologies, such as low-emissivity coatings and insulated glass units, can improve energy performance without compromising aesthetics. Brutalist structures often have limited natural ventilation due to their solid facades. Retrofitting mechanical ventilation systems while maintaining the building's distinctive appearance, in the case of the AIA HQ building or elsewhere, is a big challenge students must confront.

Efficient lighting design is essential for reducing energy consumption in Brutalist interiors. Maximizing natural daylight penetration through strategic openings or light wells can minimize the need for artificial lighting during daylight hours. Additionally, implementing energy-efficient lighting fixtures, such as LED luminaires with dimming controls, can further reduce electricity usage while complementing the building's minimalist aesthetic.

Renovating Brutalist buildings for energy efficiency should prioritize sustainable material choices. Utilizing locally sourced, recycled, or low-impact materials for insulation, finishes and fixtures can reduce the environmental footprint and enhance the building's longevity. Integrating green roofs or vertical gardens can also mitigate heat island effects and promote biodiversity while adding a contemporary touch to the Brutalist aesthetic. Assessing the life cycle environmental impact of renovation interventions is essential for holistic sustainability. Evaluating the embodied energy of materials, operational energy savings and potential carbon emissions reduction enables informed decision-making throughout the renovation process. Life cycle assessment tools can help quantify the environmental benefits of energy-efficient retrofits and guide long-term sustainability strategies.

The coursework weaves visits from the guest speakers and a field trip to the AIA HQ building to ensure the students interact with the construction team and client representatives. The field trip and assessment of the existing building is essential to the learning process of the students. The ability to ask questions of the industry experts and the client representatives of the AIA helps the students to understand they will be producing real solutions. Materials which promote health are encouraged. The assignments require smaller group work which encourages interaction, collaboration and iteration.

All of these things come to bear on the students' evaluations of the problem, as well as their proposed solutions.

Assignments

Students complete a series of small assignments pertaining to different parts of building efficiency, electrification and decarbonization to provide scaffolding for them to learn and apply key basic concepts, such as understanding the architectural features and improving energy consumption. The smaller assignments build on each other to create the foundation needed for the final project where the students collect all the design information with one goal – to decarbonation the building. The pathway to decarbonization is broken into simple steps. The first step is to create a low energy building with architectural changes. Then comes methods to electrification of the design. Next is the addition of the renewable resources and then, the final step is to decarbonize through computing the carbon and finding ways for offset. The final assignment of the course is the culmination of their learning design and application showing all the steps to achieve decarbonization.

The first assignment is to reduce energy consumption by 25% by detailing improvements to the building envelope and reducing operational carbon emission by looking at insulation, summer shading and how to improve

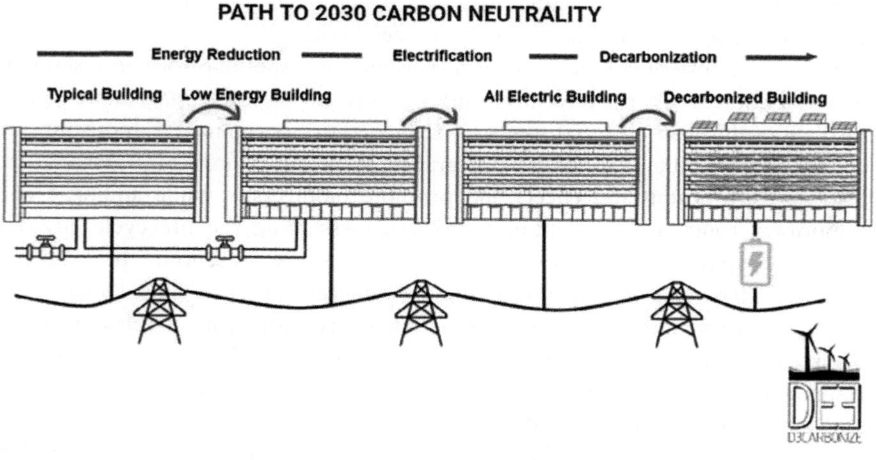

FIGURE 5.2 Example student group concept – pathway to carbon neutrality.

window panels. The next assignment is to understand different electric lighting reduction strategies and daylighting improvement strategies—along with their embodied carbon calculations using the Early Phase Integrated Carbon assessment (EPIC), created by the architecture firm EHDD, for the AIA HQ's next 50 years. Their goal is to depict the existing HVAC system and a table of proposed changes to mechanical, engineering and plumbing (MEP) equipment to electrify the building. This process of learning from existing MEP systems, researching new MEP systems and depicting new strategies, reinforces the concepts and goals of the project and, importantly, helps students understand that design interventions are systems interventions.

Another assignment asks the students to experiment with the numbers if they were to replace 100% of the cladding and 100% of the glazing, thereby increasing the embodied carbon of the building, but reducing the operational carbon. Another assignment asks the students to choose from one of several specifications for the envelope and interior fit-out and evaluate how it impacts embodied carbon. Yet another assignment asks them to estimate the EUI and operational carbon intensity of their design after their hypothetical upgrades are applied based on specific calculations (i.e. EPIC baseline = 90 kBtu/SF/yr, AIA building current EUI = 95 kBtu/SF/yr, average office EUI in climate zone = 87 kBtu/SF/yr.). Finally, students are to reevaluate their decisions and put them together with calculations of an on-site solar photovoltaic strategy (and annual estimated energy generation). The students have to implement carbon offsets, and show the baseline embodied carbon, final operational carbon and the strategy to maintain the buildings at carbon neutral design. The final project is both an oral and graphic presentation with an outside jury giving the essence of real practice in architectural firms.

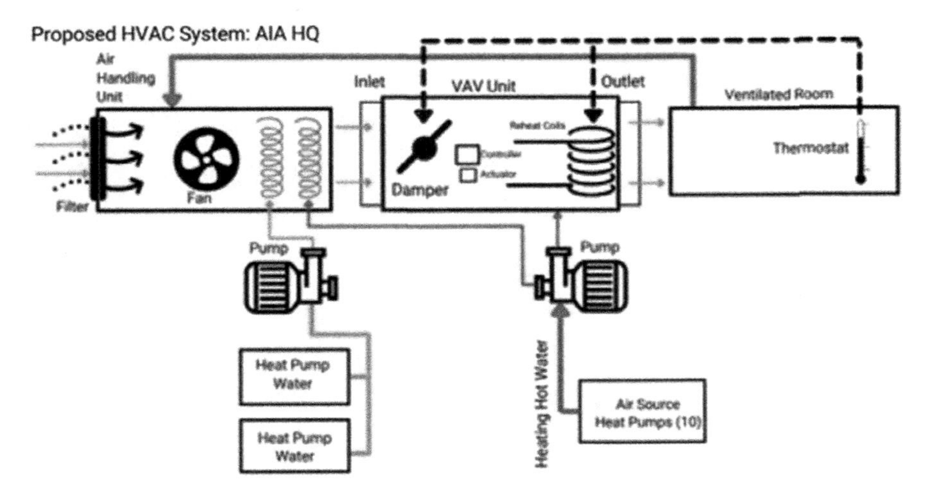

FIGURE 5.3 Example student group assignment – proposed mechanical design.

The goal of the jury is to provide thoughtful feedback on the process and practical outcome of the steps in design in fulfilling the goal of building decarbonization.

Analytics

Software is an inevitability for architecture professionals these days, and the effort to decarbonize our built environment comes with yet more tools that measure, evaluate and make recommendations such as Sefaira, EPIC, various sketching programs, and even simple calculators such as solar rebate conversions. By giving students access to these tools, especially in their final years of architecture school as fourth- and fifth-year M.Arch candidates, this course helps them also focus on the meaning of their software applications on design, in addition to learning the rules of each piece of respective software. Of course, learning new software in real-time (thereby mimicking real-world practice conditions) challenges students, and they discover the various quirks of each tool (i.e. Sefaira has lot of output to sift through compared to EPIC, while EPIC is simpler to work with to achieve accurate carbon calculations). In the end, the students prefer software that is simpler to work with, rather than data rich. The reason for that is simple: the students need to be as efficient as possible in using the software to be able to apply it as needed when iterating the design. The first course cohort of students did not know Sefaira, so it was a sharp learning curve. The EPIC software is much simpler, with default values and the 2030 carbon neutrality plug-in, which helps them gain efficiency in their work.

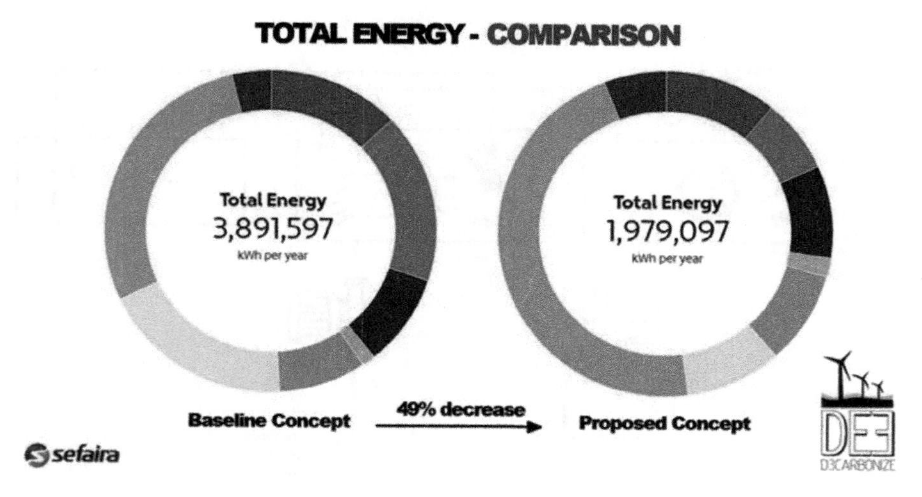

FIGURE 5.4 Example student group assignment – annual energy estimated studies.

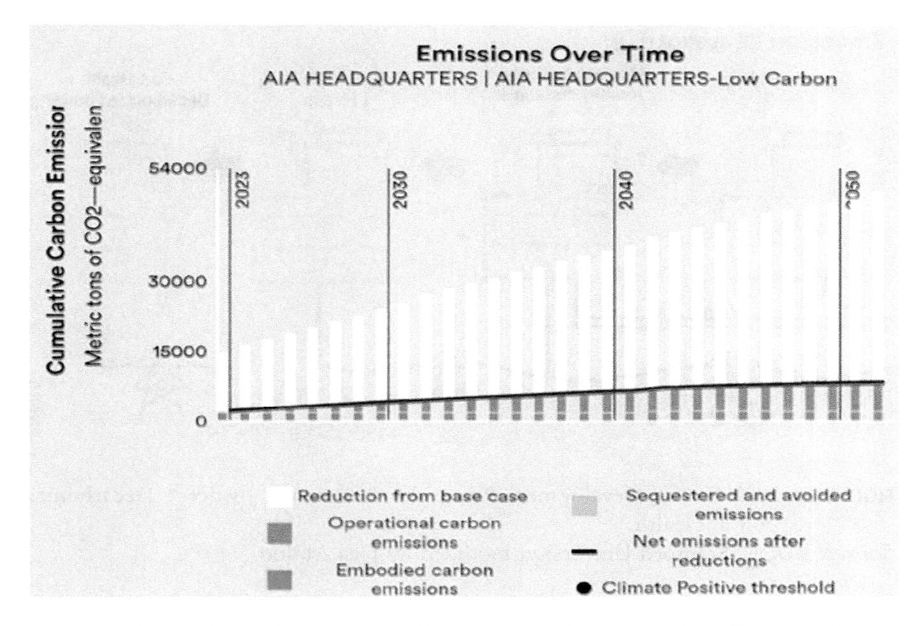

FIGURE 5.5 Example student group assignment – embodied carbon, EPIC tools.

Required Readings

Stanford University Building Decarbonization Learning Accelerator (BDLA). 2022. AIA Headquarters Renovation. May. Accessed May 1, 2024. https://bdla.stanford.edu/?post_type=case_studies&p=2922

Stanford University Building Decarbonization Learning Accelerator (BDLA). 2022. Exelixis–1951 HBP. May. Accessed May 1, 2024. https://bdla.stanford.edu/?post_type=case_studies&p=576

Recommended Reading

AIA. 2023. Framework for Design Excellence. June. Accessed December 1, 2023. https://content.aia.org/sites/default/files/2023-06/Framework_for_Design_Excellence_June_2023.pdf.

Outreach Opportunities

As this course has shaped the curriculum at Howard University and the direction towards clean energy, it has also inspired the development of the award-winning course development prize "Environmental Justice + Decarbonization + Health" from Association of Collegiate Schools of Architecture (ACSA) and Columbia University's Buell Center. It has also been the seed course for the US Department of Energy's Zero Energy Design Designation and has been covered by national CBS media. The class has been the seed for many more classes interlocking other interdisciplinary collaborations including adding affordability and scalability.

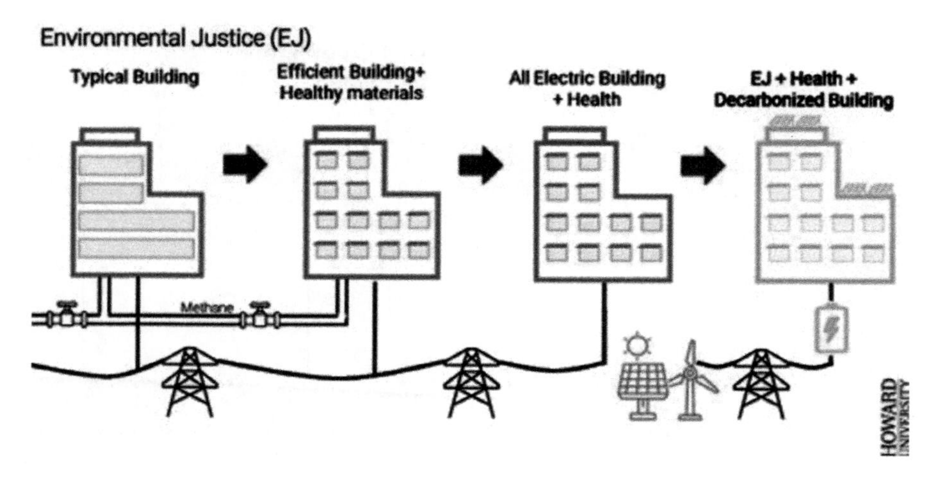

FIGURE 5.6 2022 Course Development Prize – Environmental Justice + Decarbonization + Health.
Source: BDLA, Stanford University, modified by Nea Maloo

In terms of innovation for this course, there is a pathway for future topics such as the introduction of wind energy and clean energy sources (with other forms of innovation in design).

The course has been the backbone for the formation of the instructor's first team to participate in the US Department of Energy's Solar Decathlon and, since then, a Howard University team has participated for three years in a row, implementing lessons learned from the class to practice. The Decathlon competition has robust international standards of energy metrics and carbon offset and, for that reason, has been a great canvas to test the retention of knowledge from the class and give students real practice experience.

Course Assessment

The course is designed with all the smaller assignments, group participation and class activities carried towards the final grade of the semester. Additionally, there is the final project which is a capstone project, displaying the storytelling, design and technical showcase of the student's pathway in decarbonization of the project.

Collected student feedback over the last three years will be used to inform the changes in the coursework. When asked which concepts from the course they felt were clearest, for instance, the students liked the clarity in the concepts of the MEP systems and the overall steps to achieve decarbonization of the building. This course is planned to teach the concept of how to achieve a toolkit for decarbonization for existing buildings. Students seem to have grasped that bigger goal. One student said, "The outlined path towards a decarbonized building was very useful to understand the significant aspects that need to be addressed to take on such a project." Electing to focus on other things, the course did not go as far in depth on carbon offset and pricing with the students, which a few have noted, with one saying, "The concept of carbon offset purchasing is still vague and somewhat unreachable."

The students expressed the overall value of the course including the group work, collaboration and learning new software such as Sefaira and the EPIC tool. The students' knowledge of the software, and the ease of use, either propelled or restrained the progress. The course assignments, being group work, allowed learning among peers. There was also learning with the global perspective of decarbonization and the fluidity of the course by several global speakers.

"Learning about decarbonization strategies being implemented worldwide was very useful and expanded my view of the issue of climate change. It made me consider that decarbonization strategies must really be tailored to the local climate and consider how energy is sourced for that region," noted one student. "This was indeed the most fluid course I have taken while in college," noted another.

Impact

The impact of this course is profound among the students and the faculty. The students who take the course are further able to apply the content several times in the subsequent design studios and in their final year thesis projects. In addition, they are able to apply this knowledge in their subsequent jobs in firms. The students had the opportunity to be part of the first symposium of building decarbonization held by the Stanford University BDLA. There were sessions to teach faculty on how to embed this in the curriculum. The faculty have presented at several national and international conferences.

Additional impacts include:

- Course Prize for Development, 2022. Environmental Justice + Decarbonization + Health, Howard University, Association of Collegiate Schools of Architecture (ACSA) and the Buell Center of Columbia University Course Development Prize.
- New course development, Equitable High Performance Energy Design, Howard University Zero Energy Design Designation (ZEDD) seal of recognition by the US Department of Energy's (DOE) Office of Energy Efficiency and Renewable Energy (EERE), 2022.
- CBS interview, "Howard University program focuses on reducing buildings' carbon footprint," CBS Interview, www.cbsnews.com/video/reducing-buildings-carbon-footprint-howard-university-students.

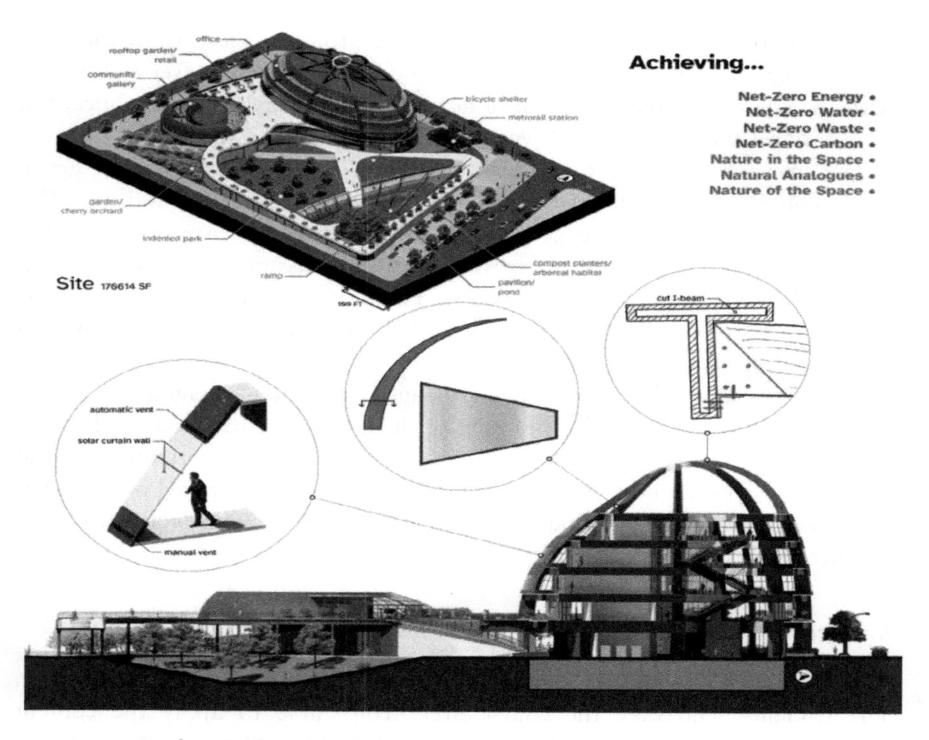

FIGURE 5.7 Student Joshua Ajayi thesis, "Zero Carbon Design."

FIGURE 5.8 Zero Energy Design Designation logo.

Looking Ahead

New courses and curriculum changes need to address the fast-changing revolution of how many people will live in cities, the impact of climate change on their lives, and the critical topic of housing affordability for all. This class has been the seed for many more classes interlocking environmental justice, affordability and scalability at Howard University, a trend that will likely continue. In terms of innovation, an "introduction to wind energy" and other forms of grid energy is essential to develop for the classroom. Similarly, the role of life cycle carbon analysis and the circular economy are other changes which need to be explored in the curriculum. The need for integrating environmental justice with decarbonization for all is important to give the students the toolkit needed to make changes in the design and implementation of buildings. There must be many more robust courses which are interdisciplinary and multi-institutional to build on the knowledge we have and to share with our students, and that's an area of innovation that must be explored, as well.

PART II
Bioclimatic Response

6

INTEGRATING CLIMATE-BASED PASSIVE DESIGN STRATEGIES INTO STUDIO PEDAGOGY TOWARDS SUSTAINABLE NET-ZERO CARBON BUILDINGS

Ulrike Passe

Course Overview

Course Name

ARCH 601: Sustainable Building Design

University / Location

Iowa State University, Ames, Iowa, USA

Targeted Student Level

- Graduate
    - 2nd year, Master of Architecture, 3-year program with advanced standing option

Course Learning Objectives

In this course students will:

- Develop projects through integrative design strategies that explore the relationship between buildings and environmental forces to maximize non-wasteful, efficient use of energy, water and materials resources.
- Investigate the impact of solar energy, airflow, materials assembly, passive and active systems on spatial quality and form creation while demonstrating synthesis of user requirements, regulatory requirements, site conditions and accessible design.

DOI: 10.4324/9781032692562-9

- Quantitatively validate design decisions through energy modeling and performance simulation.

Course Learning Outcomes

Successful completion of this course will enable students to:

- Develop architectural design projects for carbon neutral, adaptive and resilient mixed-use building projects with a focus on inner-urban housing of manageable scale that explore the relationships among architecture, cultural landscapes and environmental issues.
- Address and densify inner-urban sites and thus reduce the community contributions to carbon emissions.
- Conduct interdisciplinary research.
- Engage with local stakeholders.
- Respond to a selected climate and site conditions, which changes every year.

Course Methodology

Philosophy

> Through exclusively social contracts, we have abandoned the bond that connect us to the world, the one that binds the time passing and flowing to the weather outside, the bond that relates the social sciences to the sciences of the universe, history to geography, law to nature, politics to physics ... We can no longer neglect this bond.[1]

The impact of buildings on the changing climate and resource depletion leads to the demand for carbon neutral buildings as one major aspect of the 17 UN Sustainable Development Goals.[2] A rapid reduction in carbon emissions (both embodied and operational) is needed to achieve the Paris Agreement's goal to limit global warming. However, buildings are complex socio-technical entities, embedded in a social-cultural-economic-climatic context, and carbon or energy cannot be considered in isolation. Therefore, carbon neutral buildings need to be embedded within a holistic education towards sustainable architecture and cities. It is thus imperative to design buildings at the intersection of the social and the natural contract as proposed by Michel Serres (above).

Design pedagogy relies heavily on "learning by doing."[3] Designing is learned by mocking-up a virtual real-world project. Therefore, ARCH 601 directly integrates performance simulation of the technical predicted operation into the workflow so that the simulations drive the process.

The urgency to change how we construct our built environment demands integration and collaboration of all stakeholders in the design and construction process. Given these critical demands for sustainable design, architecture technology and design pedagogy also need to be integrated. Form and performance need to be developed based on a holistic conceptual understanding of space and materials with all their complex environmental, phenomenological and sensual properties. Visualization and awareness of all sensual qualities of space is the crucial starting point toward providing a holistic architectural education.

In ARCH 601 Design and Environmental Forces, Systems and Controls are intrinsically related through the materialized relationship between built form and history, culture and the environment, in a particular climate, location and topography. Historically, humidity, aridity, heat, cold and light intensity have left their marks on the roof types, surfaces, materials and incorporated openings of structures. They are positioned in relationship to the direction of the sun, the prevailing winds and temperatures. The concepts for heating, cooling and lighting are based on the interrelationship of the exterior climate and internal needs and are thus strongly related to design principles and spatial practices. This critical inquiry is based on many (sometimes conflicting) parameters. New digital tools allow the integration of energy and environmental performance, so the potential for parametric modeling has become even more complex, but also visually more accessible to designers. Design is an iterative process of thought and creation that integrates various parameters and forces them into an inextricable relationship of space and form.

Physical relationships are expressed in mathematical formulae translated into architectural and spatial proportions, phenomena and sensations. This approach leads to an interrelationship between geometry, material, space, color and atmospheric conditions. Qualities should be stressed rather than mere calculations and pure physical quantifications. Concepts of physics enable ARCH 601 students to communicate performance metrics with their engineering collaborators in their future professional lives. Using state of the art research findings in the humanities, the sciences and engineering, integrative learning experiences guide students to become architectural professionals, researchers, or scholars with knowledge of sustainable design concepts, theories and practices.

Definition

Carbon neutrality – Net-zero energy buildings with low embodied carbon material selections.

Implementation

An emphasis is placed on a regional site, and its socio-economic and historical conditions, which is representative of similar situations throughout the Midwest and South of the US mirroring the development of the US American society in its global context. The projects stress interdisciplinary research, engagement with local stakeholders and contemporary phenomena as well as response to the specific climate. Environmental stewardship guides the development of an integrative project proposal. ARCH 601 addresses NAAB student performance criteria SC.5 Design Synthesis and students work individually at first and then in small teams of two.

The Role of Precedents

Students gain thorough understanding of how spatial and programmatic design concepts relate to energy performance by conducting a post occupancy evaluation of a precedent project while they develop performance simulation skills. All precedent projects are studied with simulation tools introducing the key metrics, such as energy use intensity (EUI)[4] daylight autonomy (DA),[5] natural ventilation potential, and the goals of the Architecture 2030 Challenge to reduce demand. Passive strategies are prioritized to reduce the loads of active and renewable systems.

Site Context: The Community Dimension

Sustainable development in architecture and urban development (UN SDG 11) demands the integration of societal issues with the environmental and economic. In ARCH 601 the social is introduced via deliberate site selection. From 2014 to 2023 sites were carefully selected to include considerations of the impact of place on social dimension of sustainability in Old North St. Louis (ONSL), Missouri; Sixth Avenue Cultural Corridor in Des Moines, and Greene Square Downtown Cedar Rapids, Iowa; 324 South Front Street, Midtown Memphis, Tennessee; and Pershing Crossing, City of Chicago, Illinois. Three "Lost Iowa Schools" were used to develop viable adaptive reuse project towards carbon neutrality in 2016, and Old South Baton Rouge Louisiana provided sites in 2023. A thorough study of the site's socio-economic and demographic context provides the students with the basis to develop their own program. Project sites are located either in an urban context which provides challenges to the ideal relationship to solar orientation or are adaptive reuse projects. In all locations, the students are charged with the development of a mixed-use program combining residential use, an environment for working and spaces for community, which address the socio-economic situation through density.

Workflow to Facilitate an Iterative Design Process for Decision-Making

The conceptual design workflow was tested and developed over 10 years (Figure 6.1). It comprises three phases starting with climate and radiation analysis leading to decisions on orientation, massing and geometry, followed by an iterative process which includes daylight and glare analysis, thermal energy balance, natural ventilation and spatial composition as well as early integration of a detail interlude. This interlude strengthens the students' understanding of material and assembly processes in relationship to data inputs in the modeling tools. Iterations can be repeated till a final synthesis is developed.

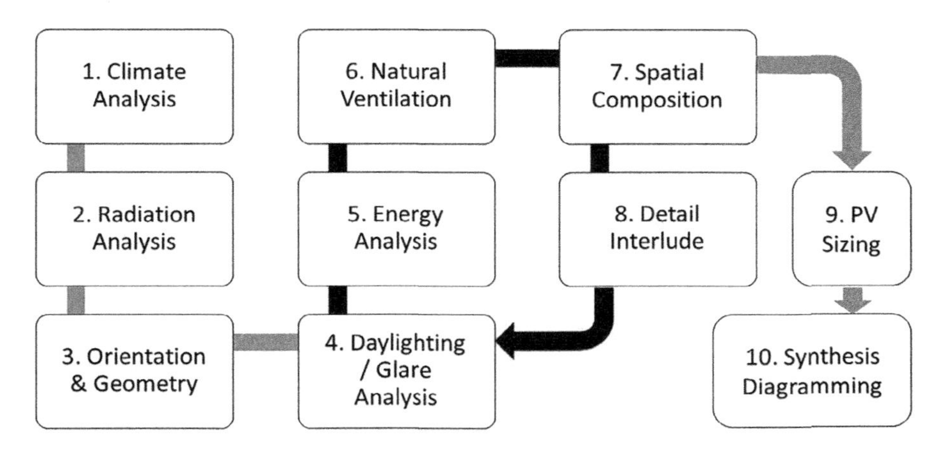

FIGURE 6.1 Workflow diagram showing the iterative sequence of the ten stages.

Stages 1–2: Climate and Radiation Analysis

To develop volumetric concepts, the teams study the relationship between space and environmental forces. Depending on their location, energy performance of buildings is based on their potential to utilize solar radiation on site in winter and sheltering the interior from solar radiation in summer while still providing sufficient daylighting and ventilation. Seasons shift with location and climate and ARCH 601 utilized a different location every year. At least three volumetric parti diagrams evaluate solar radiation, daylighting and ventilation strategies for apertures to optimize orientation. During the warmest period of the year reducing solar radiation needs to balance daylighting, while during winter, maximizing passive solar balances heat loss during the cold or very cold nights. Passive cooling in summer is as important as passive heating in winter yet the durations change with location. Natural ventilation potential largely relates to the relationship of the building's position and orientation to the prevailing winds in summer and sheltering the building or group of buildings from those same forces in winter. The rarely well-understood concept of stack ventilation is introduced at this point.[6] A good ventilation strategy requires the connection of vents via a flow path. This stage highlights the importance of a shallow floor plate, optimizing the relationship of spatial composition to the prevailing wind directions and creates space and form to channel and block air. Climate Consultant[7] (Figure 6.2) provides passive design strategies to explore in the next steps with refined computational tools.

Stage 3: Developing Massing Geometry and Orientation to Reduce Energy Loads

Since 2020 massing studies resulting from Stage 1 and 2 are then tested with ClimateStudio for Rhino for their potential annual energy use intensity (EUI) to

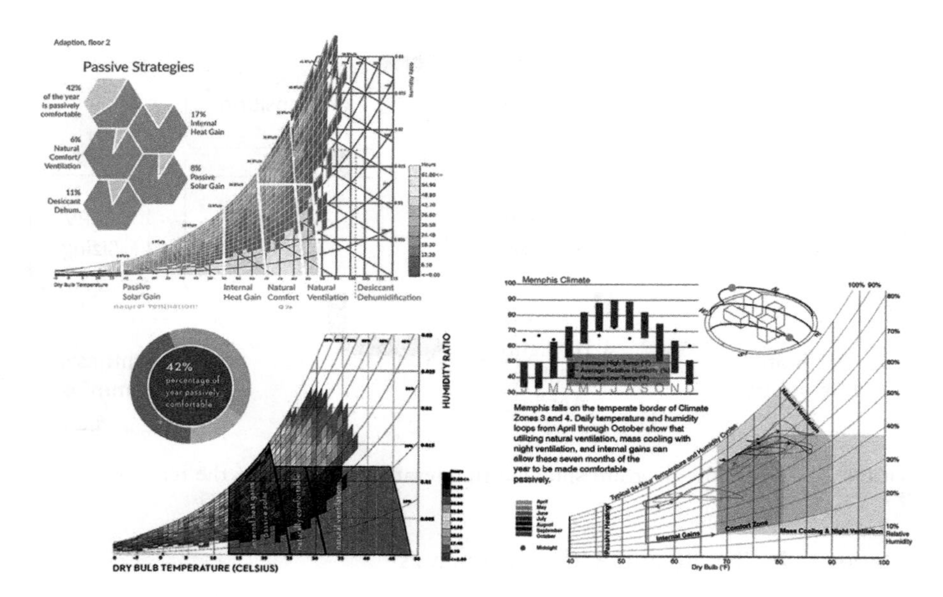

FIGURE 6.2 Psychrometric climate analysis for Ogden, Iowa (2016), St. Louis, Missouri (2014) and Memphis, Tennessee (2019).

meet the Architecture 2030 Challenge (2023) goals.[8] Previous cohorts used DIVA4Rhino and Sefaira.[9] Keeping program variables constant at first, the most promising massing schemes are iteratively tested with a variety of window to wall ratios and enclosure strategies. Teams then move forward with at least two partis in parallel (Figure 6.3).

Stages 4–5: Energy Load Reduction vs Daylight and Views

Solar radiation and parametric daylighting tools previously in DIVA now within ClimateStudio for Rhino[10] are used in this two-week session to determine window to wall ratio, extend the time of year when no active systems are

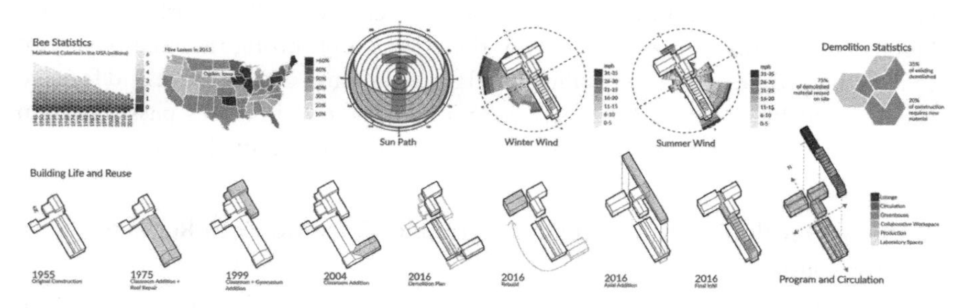

FIGURE 6.3 Massing studies for the Apicenter, Ogden Iowa School project by Shawn Barron and Saranya Panchaseelan (2016).

needed and to integrate daylight autonomy targets and natural ventilation strategies into the overall spatial layout (Figure 6.4).

Stages 6–7: Ventilation vs Energy vs Daylight to Form a Spatial Sequence

Introducing natural ventilation is often the largest challenge for the emerging sustainability designer as the resource to facilitate natural ventilation (wind) is erratic and volatile, especially in an urban context. Yet it provides the most opportunities for spatial exploration. Opening windows in the climates along the Mississippi River Basin are often not sufficient to provide the required air change rate and cooling capacity. Thus, during these stages the students alternate iteratively between daylight simulation for 50% autonomy and natural ventilation studies comparing flow path, window to wall-ratio and aperture strategies to achieve the most possible hours for natural ventilation per year. MIT Coolvent[11] used to be a robust investigation tool because it is quick and provides sufficient information for informed decision making. The java-based computational tool allows input for spatial schematics based on five sectional

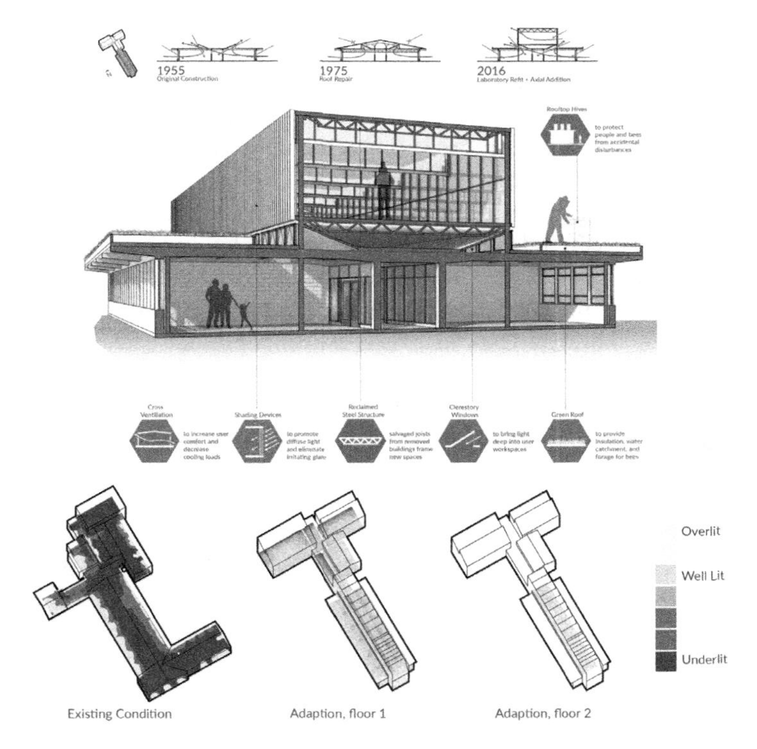

FIGURE 6.4 Apicenter studies detailed spatial daylight autonomy in an adaptive reuse project which revives the original composition of the school building and adds new volumes on the roof (2016).

typologies (atrium, cross-, single-sided-, shaft and chimney) and provided sufficient results. A new tool still needs to be found.

Stage 8: Detail Interlude: Material Consideration for Carbon Neutrality and Beauty

Architectural design concepts manifest themselves in the way the building is constructed and detailed by separating, layering or joining architectural elements according to the intended spatial and formal composition and according to technical necessities to achieve a functional whole. This conceptual notion gets elevated to a new level of complexity when energy efficiency needs implementation into the actual construction of a building. Therefore, half-way through the 16-week semester, after the schematic design outcome of their iterative process from stages 3 and 7, a weeklong exercise challenges the students to develop a large-scale detailed section and respective elevation at half-inch to a foot scale (equals roughly 1:25) following the path of the sun and air in summer and winter.

Students engage in a critical discourse on the need to design more airtight buildings with high R-values and high-performance windows reconciling appearance and aesthetic with environmental, building physics and structural concepts. These drawings / models represent a space and its materiality and atmosphere; and are not merely representing technical assembly. A variety of material choices are discussed based on their embodied carbon assessment using available carbon calculator EC3.[12]

The Detail Interlude also requires the students to consider urban – landscape integration and access related to the site topography and thus the integration into the site and urban context while developing spatial connections from urban exterior to private interior. The detailed sectional drawings are red-lined and discussed in collaborative sessions, where students provide each other feedback. The Detail Interlude provides the basis for the most important final integrative section drawing (Figure 6.5).

Stage 9: Meeting Reduced Energy Demand with Renewables

Once the students have refined their design proposal and reduced the required electrical energy demand significantly via massing, envelope, daylight, solar heating and passive cooling strategies, as well as embodied carbon towards a net-zero carbon potential in stage 1 to 8, the photovoltaic (PV) array is sized on and around their built volumes. The opportunities for using PV for previously developed shading strategies and window integrated technologies are explored as well to synergistically use shading elements which reduce heat gain and produce electrical energy. The placement and integration of the renewable energy source is treated as a major design integration challenge using the PV tool within ClimateStudio for Rhino or basic hand calculations. All prior

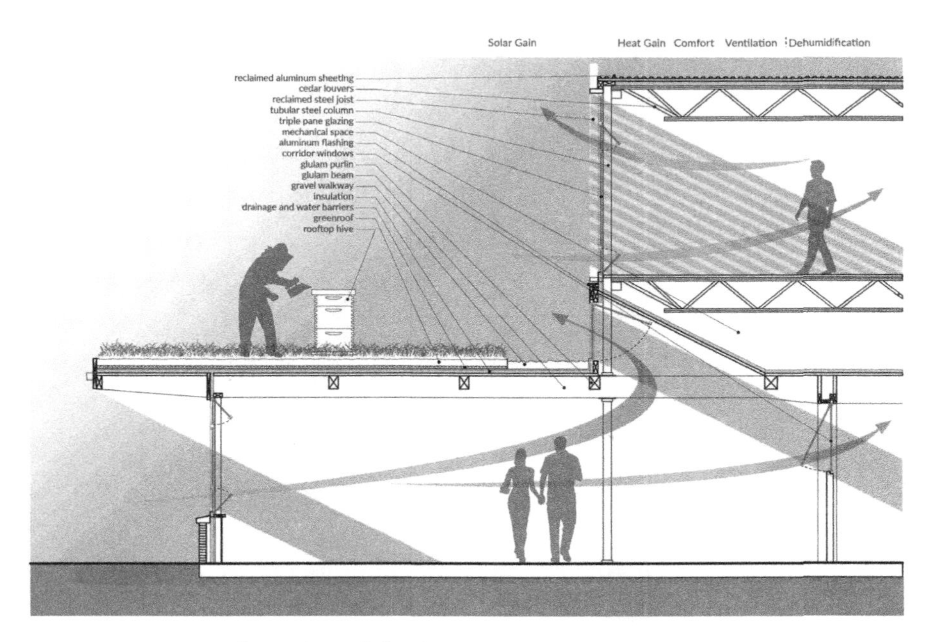

FIGURE 6.5 Large scale section of the Apicenter (2016).

design decisions have contributed to reducing the energy use, which needs to be produced by the photovoltaic array. Achieving a net-zero energy balance is the first step to carbon neutrality, the second is the consideration of embodied carbon (Figure 6.6).

Stage 10: Final Design Synthesis and Diagramming

After multiple iterations all performative aspects are integrated into a built form and correlated with the desired building program based on a critical reflection of envelope material and construction type, with determination of window to wall and floor area ratio to improve understanding of the depth of enclosure assemblies selected in the energy simulation tool. The goal for a low or zero carbon emission building is thus also an exercise in composition, space planning and aesthetics integrating exterior and semi-exterior spaces (Figure 6.7).

Analytics: Computer Models in Architecture Pedagogy

Performance based design decision scenarios provide insights into the future operation of the building when the success of a design strategy relies on the proper control of air movement, daylight and sunlight penetration or blockage leading to the control of energy flow in the building. As the project does not exist yet, the iterative process analyzes the opportunities of the site. The

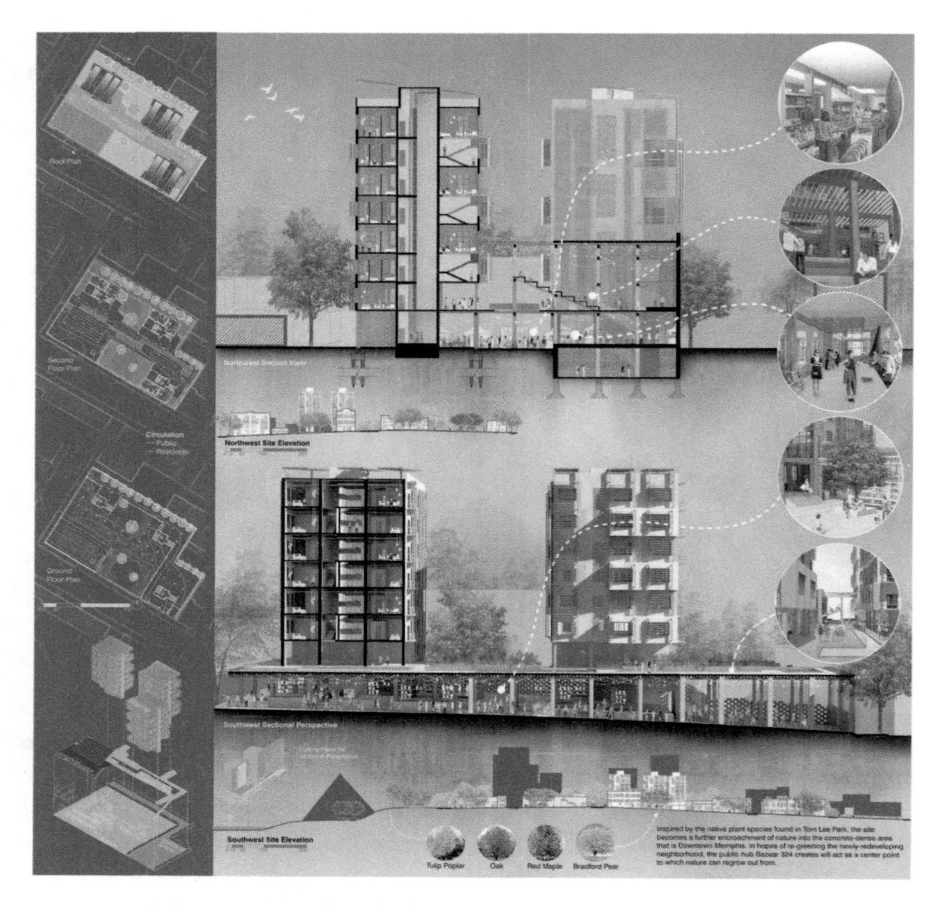

FIGURE 6.6 Full competition board of Bazaar 324 for Memphis TN (2019) by Connor Mougin and Anannya Das.

studio follows the definition by Jankovic (2012) towards computer based performative design:

> the term modeling is defined as making a logic machine, which represents the geometry and material properties of the building and the physics processes in it. Simulation is then defined as numerical experimentation with the model so as to investigate its response to changing conditions inside and outside the building.[13]

In the same way as physical models, digital computer models require a significant degree of abstraction. Not all geometric details are necessary and not all physics processes can be simulated due to time limitation. This is an important learning objective of the studio which often goes unnoticed when working with digital tools. Models are abstract representations of an aspect of the world as numerical equations are used to simulate physics in the real world.[14]

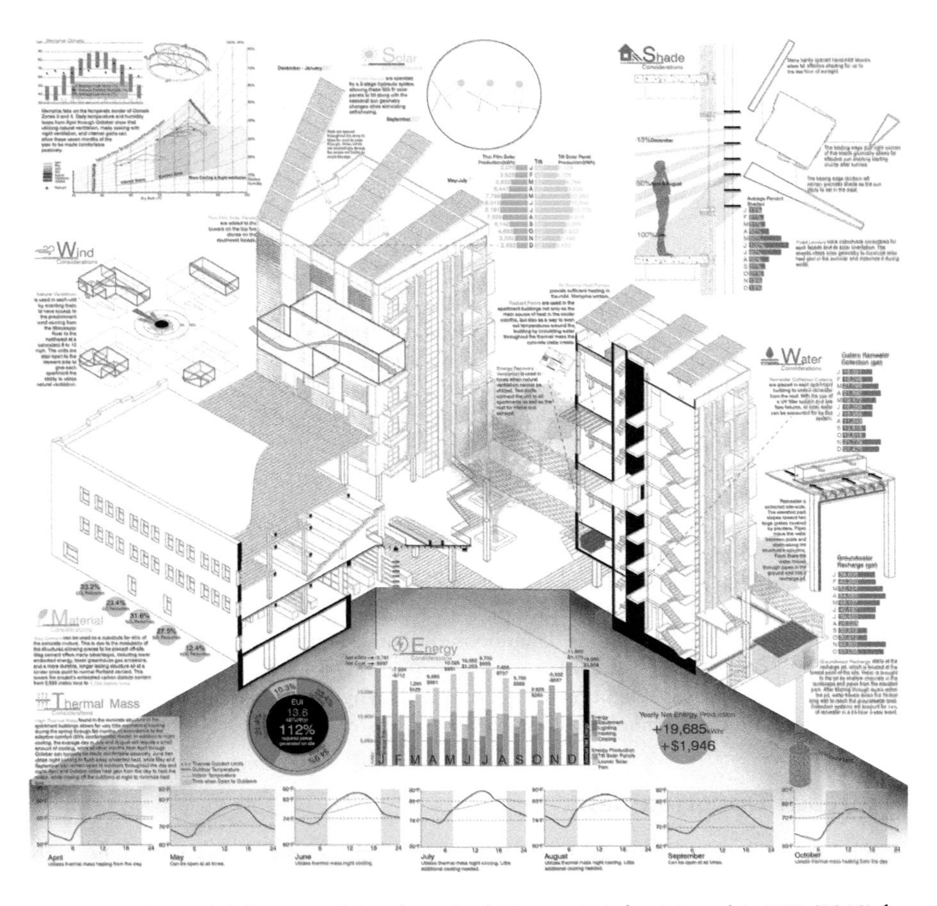

FIGURE 6.7 Second full competition board of Bazaar 324 for Memphis TN (2019) by Connor Mougin and Anannya Das.

The account of climatic variability in performance-based design depends on the science behind computer simulation. Thus, the students make decisions based on performance experimentation conducted using the computational model. To understand the relationship between the mathematical inputs for physical relationships into the computer model all students in the course are required to first take the Environmental Forces and Control Systems module of the required Building Technology sequence, where they conduct the calculations by hand and learn to understand the impact of each variable on the overall outcome.

Required Readings

Hausladen, Gerhard, Michael de Saldanha & Petra Liedl. *Climateskin: Concepts for Building Skin that Can Do More with Less Energy*. Birkhaeuser, 2008.

Jankovic, Ljubomir. *Designing Zero Carbon Buildings Using Dynamic Simulation Methods*. Earthscan from Routledge, 2012.

Kwok, Alison & Walter Grondzik. *The Green Studio Handbook*, 3rd Edition. Architectural Press, 2018.

Recommended Readings

Deplazes, Andrea (ed). *Constructing Architecture: Materials, Processes, Structures: A Handbook*, 3rd Edition. Springer, 2013.

Grondzik, Walter, Alison Kwok, Benjamin Stein & John S. Reynolds. *Mechanical and Electrical Equipment for Buildings*, 13th Edition. John Wiley & Sons, 2014.

Hausladen, Gerhard et al. *Climate Design: Solutions for Buildings that Can Do More with Less Technology*. Birkhaeuser, 2005.

Hemsath, Timothy & Kaveh Alagheh Bandhosseini. *Energy Modeling in Architectural Design*. Routledge, 2018.

Hensen, Jan L.M. & Roberto Lamberts (Editors). *Building Performance Simulation for Design and Operation*, 2nd Edition. Routledge, 2019.

Lechner, Norbert. *Heating, Cooling, Lighting: Design Methods for Architects*, 4th Edition. John Wiley & Sons, 2014.

Passe, Ulrike & Francine Battaglia. *Designing Spaces for Natural Ventilation: An Architect's Guide*. Routledge, 2015.

Reinhart, Christoph. Daylighting Handbook 1: Fundamentals Designing with the Sun. Building Technology Press, 2014.

Reinhart, Christoph. Daylighting Handbook 2: Daylighting Simulations and Dynamic Facades. Building Technology Press, 2018.

Outreach Opportunities

Each studio developed professional and community connections with whom the students interacted during the site visits.

- 2014: Old North St. Louis Neighborhood Association, St. Louis, Missouri.
- 2015: Architecture firms and community organizations, Des Moines, Iowa.
- 2016: ISU Extension and various Iowa School Districts.
- 2019: Memphis Design Collaborative and Downtown Economic Development Group.
- 2020: Durham North Carolina Haven Ventures.
- 2021: Sabathani Community Center, Minneapolis, Minnesota.
- 2023: Kimble Properties, Baton Rouge, Louisiana.

Student teams also achieved competition successes with the following AIA COTE Top Ten for Students Competition winners:

- 2014/2015: Re-Sustaining Old North Saint Louis, Kyle Vansice, Brandon Fettes; HRR: Harvest, Recycle, Reuse, Heidi Reburn, Sean Wittmeyer; and Old North Bikes St. Louis, Stephen Danielson, Benjamin Kruse.
- 2015/2016: The Living Link, Mengwei Liu, Anastasia Sysoeva.

- 2016/2017: The Apicenter, Shawn Barron, Saranya Panchaseelan.
- 2019/2020: Bazaar 324, Connor Mougin, Anannya Das.

Course Assessment

Graduate students from 2014 to 2018 (39%) and 2019 to 2023 (30%) responded to an Institutional Review Board approved anonymous survey about the importance the knowledge gained in ARCH 601 has in their current professional career. Figure 6.8 highlights the main quantitative responses. Only 50% are still using performance software in their current career. The numbers in the first and second group were fairly similar, thus they are combined here.

Qualitative reactions provided further feedback:

> In ARCH 601, we integrated sustainable practices that allowed us to further explore design options with sustainability and general well-being was at the forefront. It also allowed us to very early on in our education apply comprehensive systems in order to clearly understand the relationships between the variety of systems at play. It also allowed us to play with parametric analysis tools that allowed us to visualize the climate impacts clearly.
>
> Learning and applying the metrics for net-zero design, including passive orientation, daylighting, envelope analysis, and connecting the model with construction technology was invaluable.

Looking Ahead

Computational energy simulation tools are constantly evolving, and with the release of ClimateStudio by Solemma in 2020,[15] computation time for daylight and glare simulations were significantly reduced enabling more detailed focus on analysis than on simulation runs. Secondly, while operational carbon assessment is

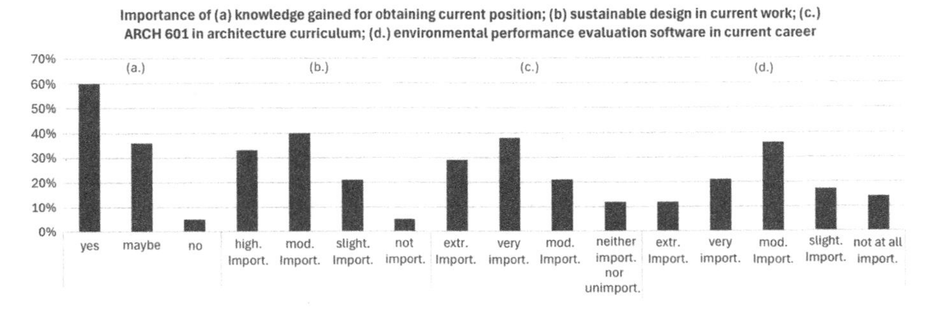

FIGURE 6.8 ARCH 601 graduate student responses on the importance of sustainable design 2014–2023.

currently well understood, embodied carbon tools are still challenging for students to use, thus future developments in this area will significantly improve trust in those results. Lastly, future studios will continue to focus on midsize urban settings, adaptive reuse also in suburban areas such as abandoned warehouses and malls will become a topic. Site water issues and site development will provide opportunities for further innovation in future studios.

Acknowledgment

This chapter is adapted from Passe, Ulrike, 2020, "A Design Workflow for Integrating Performance into Architectural Education", *Buildings and Cities*, 1 (1), 565–578, https://doi.org/10.5334/bc.48. I reorganized the material based on the required format and focus of this book, added new student work from 2019 and data from a second survey (2020 to 2023). I thank all former students for their active participation.

Notes

1 Michel Serres, 1995, *The Natural Contract*, The University of Michigan Press, p. 48.
2 United Nations, Sustainable Development Goals, https://sdgs.un.org/goals, accessed December 9, 2024.
3 John Dewey, 1916, *Democracy and Education; An Introduction to the Philosophy of Education*, The Macmillan Company.
4 EnergyStar, EUI Target Finder, https://www.energystar.gov/buildings/tools-and-resources/target-finder, accessed December 9, 2024.
5 C. F. Reinhart, J. Mardaljevic & Z. Rogers, 2006, Dynamic Daylight Performance Metrics for Sustainable Building Design, *Leukos*, 3(1), 7–31.
6 Ulrike Passe & Francine Battaglia, 2015, *Designing Spaces for Natural Ventilation: An Architect's Guide*, Routledge.
7 Climate Consultant, www.sbse.org/resources/climate-consultant, accessed November 24 2023.
8 Architecture 2030, New Buildings: Embodied Carbon, https://architecture2030.org/new-buildings-embodied/, accessed December 9, 2024.
9 Ulrike Passe, 2020, A Design Workflow for Integrating Performance into Architectural Education, *Buildings and Cities*, 1(1), 565–578, https://doi.org/10.5334/bc.48
10 Solemma, ClimateStudio for Rhino, www.solemma.com/climatestudio, accessed November 22, 2023.
11 MIT Coolvent, http://coolvent.mit.edu/, accessed February 1, 2020.
12 Building Transparency, 2019, Embodied Carbon in Construction Calculator EC3, www.buildingtransparency.org/en/, accessed December 9, 2024.
13 Ljubomir Jankovic, 2012, *Designing Zero Carbon Buildings Using Dynamic Simulation Methods*, Earthscan from Routledge, p. 7.
14 Eric Winsberg, 2010, *Science in the Age of Computer Simulation*, University of Chicago Press.
15 Solemma, ClimateStudio for Rhino, www.solemma.com/climatestudio, accessed November 22, 2023.

7

THE BIOCLIMATIC DESIGN STUDIO

Dorit Aviv and William W. Braham

Course Overview

Course Names

ARCH 7080: Bioclimatic Design Studio with ARCH 7540: Performance Design Workshop

University / Location

University of Pennsylvania / Philadelphia, PA, USA

Targeted Student Level

- Graduate
    - 1st year, 2nd semester, post-professional Master of Environmental Building Design (MEBD)
    - 1st year, 2nd semester, post-professional Master of Science in Design with a concentration in Environmental Building Design (MSD-EBD)

Course Learning Objectives

In this course students will:

- Understand and apply the principles of bioclimatic design.
- Create a testable, measurable design hypothesis about a design intervention.
- Select and apply appropriate performance analysis tools and methods to evaluate building design proposals.

DOI: 10.4324/9781032692562-10

Course Learning Outcomes

Successful completion of this course will enable students to:

- Analyze local climate and available environmental resources.
- Develop a project with a high level of detail, and delineate it through analysis, drawing, modeling, simulation and physical prototyping.
- Develop a language of representation for the project which demonstrates both its tactile visual and material qualities and its ephemeral qualities such as thermal properties, light and airflow.

Course Methodology

Philosophy

The primary objective of the studio is to explore what it means to *Design with Climate*, to create an architecture that harnesses environmental forces towards its own performance as a shelter for human inhabitation on a specific site. By optimizing for climate and comfort, projects naturally focus on architectural forms and materials, and lead to buildings with very low energy demand. Energy demand is further reduced by designing systems that draw on available environmental resources, such as solar insolation and wind, and utilizing locally available temperature sinks and sources, such as evaporation, sky radiation and ground temperatures.

Enhancing the bioclimatic agency of building enclosures remains the foundation of any approach to environmental modification and provides valuable elements of architectural expression. The studio allows students to understand that regions across the globe share certain key characteristics, and to learn from a variety of architectural responses to those characteristics.

Definitions

Carbon neutrality – Not an explicit term of discussion in the studio. The emphasis was on bioclimatic agency to achieve low-energy buildings and to use available environmental resources to satisfy the remaining energy demands.

Bioclimatic agency – The capacity of building designs to work with the resources and energy gradients in local microclimates to achieve healthy, comfortable environments to support human activities.

Implementation

The *Bioclimatic Design Studio* and its two parallel courses, the *Performance Design Workshop* and the *History and Theory of Architecture and Climate*, evolved together over a half dozen years at the beginning of the EBD program.

What began as a net-zero energy studio shifted to focus on design for thermal comfort and climatic adaptation. Design for net-zero energy usually proceeds typologically by benchmarking normative buildings and focusing on efficiency improvements. In contrast, design for thermal comfort and climate begins by exploring the thermal effects of architectural elements in a specific climate, which allows for more architectural exploration and results in radically low-energy designs. Using comfort as a performance metric also leads to the exploration of more thermodynamic mechanisms of heat and energy transfer and demands more inventive methods of research and analysis. Taken together, we believe this better prepares students to design low energy, low carbon buildings.

As a starting point, students are asked to investigate the forces that constitute climate. Olgyay's paradigm of the four climatic zones is a helpful first step at characterizing the interaction between architecture and the environment.[1] It helps students to understand that regions across the globe share certain key characteristics. The course on the *History and Theory of Architecture and Climate* (7180) introduces students to the global variety of architectural responses. The regional understanding of climate must further be enhanced with analysis of climate in urban settings that include the effects of human-made environments on thermal comfort.

The city is a territory of diverse microclimates, each a site for thermodynamic interactions where energy flows are in constant exchange with human-made structures. Through these interactions, the magnitudes of forces such as radiation and convection can be enhanced and multiplied or, conversely, suppressed dramatically. A specific urban microclimate is a dynamically reciprocal action between naturally occurring global cycles and their interface with urban canyons and skyscrapers, concrete roads and reflective-glass curtain walls. Any design decision will affect this energy field, so we seek to redefine the architectural design procedure in the city. First, the students use dynamic software tools and physical modeling to map the forces affecting a site. Second, they test interventions in these newly mapped and visualized energy flows through architectural design proposals.

Program

Students receive a common program, while each individual or group of students selects a different site located in a different climatic zone. The class collectively explores how the architecture of a certain program is transformed from one site to another based on climate and context. Over the 10 years of teaching the bioclimatic studio, we have used different programs as case studies for design exploration from mixed-use buildings to schools and data centers. Below are examples of student work from three years of the program.

Bioclimatic Studio 2018: A New Chautauqua Institute

The program for the studio was the New Chautauqua Institute, an international research and development corporation, leading the transition to a prosperous, renewable economy. The mixed used building of about 50,000 sf was chosen to allow the coordination of programs with different schedules and energy requirements. It combined facilities for outreach and education, mixing the typologies of office, university, conference hotel and business incubator. The facilities also had to adapt as the Institute's mission evolved and the cities, economies, or climates in which they are located changed.

The method of the studio was bioclimatic hybridization, a building that keeps itself comfortable without explicit conditioning equipment, but with responsive modification of its envelope. The bioclimatic analysis was paralleled by a life-cycle analysis, using emergy (with an "m") synthesis, to evaluate the trade-off between materials of construction and the energies of operation, and the advantages of adaptation over time.

Contemporary buildings are inherently hybrid, combining the traditional, bioclimatic elements of buildings– walls, windows, doors– with increasingly intelligent technologies for delivering modern services. These systems of power and control have mostly been used to compensate for the inadequacies of building envelopes, but when they are successfully hybridized, innovative and powerful new forms of building emerge.

Bioclimatic Studio 2020: Reimagining the Urban Data Center

The data center is an emerging building typology, proliferating in the 21st century to provide cloud computing services to a rapidly growing number of users worldwide. Data centers are the physical infrastructure of all cloud computing applications. These centers consume large amounts of energy, partially due to the servers' electricity load, but more significantly, 40% of the energy load is spent by the mechanical systems necessary to reject the heat produced by the servers. In order to minimize the environmental footprint of data centers, the studio sought architectural strategies to mitigate that enormous heat load. Is there a model data center that uses that power to contribute to its environment? Can we envision an architecture that provides cloud infrastructure while managing its environmental impact?

The design studio explored the potentials of climate-adaptive buildings for data centers in the different climatic zones in the United States: temperate, cold, marine, hot-humid and hot-dry. During a studio trip to Seattle the students were able to visit existing data centers, to meet with industry professionals from Microsoft and Google, and to learn about the requirements and challenges involved in data center design.

Programmatically, the data center presents a fascinating building type for architects, because the majority of the building's volume is dedicated to sheltering and providing the thermal needs of machines rather than of human beings. As both inhabit the contemporary city, a synergy between servers and humans must be explored. Studying the potential thermodynamic cycles and heat transfer processes between human and machine spaces for energy efficiency became part of the design research. Daylighting opportunities and the overall experience of inhabiting the data centers played a role in the design as well. Data centers are becoming more common in urban areas across the planet, so the studio sought to develop a new building typology for an urban, climate-responsive data center.

Bioclimatic Studio 2021: A Naturally Ventilated School

During the COVID-19 pandemic, school buildings were shut down because of the risk of indoor transmission of the SARS-CoV-2 virus. In educational buildings that remained open, ventilation rates had to be significantly increased, resulting in energy consumption spikes. Having witnessed this global environmental health crisis, we had an obligation and an opportunity to rethink school design to provide both energy-responsible and healthy educational settings.

Open-air schools have a long history related to pandemics, dating back to the beginning of the 20th century. However, in climatic zones where the exterior conditions are far from the comfort zone, how do we provide enough fresh air without expending huge amounts of energy? Naturally ventilated schools are common in many countries and provided important case studies.

Wind and buoyancy are the two natural ventilation forces available to designers. The class studied how forms and materials can be used to enhance these forces and provide fresh air and cooling to interior spaces. In the *Performance Design Workshop* students explored different ventilation strategies and built scaled physical prototypes to validate and visualize air flow.

The 2021 Bioclimatic Design Studio reimagined school design in different climatic zones in relation to health, energy, environment and education. The studio explored the potential to create a climate-responsive building model in three different climatic zones: temperate, hot-humid and hot-dry.

Assignments

Phase I: Climate and Microclimate Analysis

In the first phase, students work in groups to analyze the climatic characteristics of selected sites within the four climate zones, including both simulation-based studies and GIS data analysis. The simulation-based studies used the parametric capabilities of Grasshopper to test the effect of different building elements on the thermal and luminous behavior of a simplified, equivalent “shoebox” model, sometimes rising to thousands of variations, to understand what combinations of elements are effective in different climates (Figure 7.1). The analysis uses available weather files for temperature and humidity measurements, CFD wind simulation and solar gain simulation. For the urban sites, a simple wind-tunnel is used to map ground level wind effects, complemented by CFD studies of particular conditions.

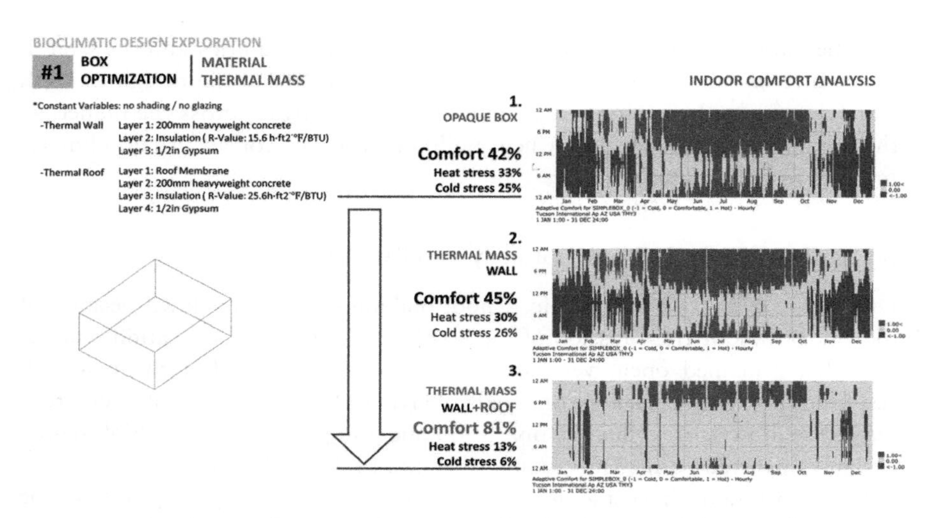

FIGURE 7.1 Bioclimatic design exploration with "shoebox" model. Parasol, Tucson. Source: Youngjin Hwang, Silmi Farah, Yunqian Li

Phase II: Programmatic Analysis, Massing + Bioclimatic Design Strategies

In phase two, student teams study the opportunities of the building program, developing an understanding of the necessary building area and volume based on indoor and outdoor activities, classrooms and support areas. In parallel, specific bioclimatic strategies for the climate are explored along with their intersection with the program and the architectural narratives they inspire.

Phase III: Design Development

With both the climatic and programmatic parameters as guidelines, teams develop design schemes appropriate to their specific site. The design studies include comparison of different form and massing configurations, building material options, and envelope design strategies for natural ventilation and self-shading. Each iteration in the refinement of the design is accompanied by evaluation of the thermal conditions and energy load. The overarching goal was to reduce the energy load on the building's mechanical system and to optimize bioclimatic techniques and natural ventilation.

Analytics

The Bioclimatic Studio follows an intense first semester in which students take required courses designed to provide a foundation for the studio. Those include a course on the fundamentals of building science, a course in *Building Performance Simulation*, and a third course that teaches a comprehensive, emergy-based method of environmental accounting.

The studio serves as a synthesizing moment for the many skills taught in those courses. We believe that architects only really learn to use performance analysis when they are trying to answer design questions, so the studio provides an opportunity to learn about the limits of simulation tools and the importance of workflow.

The tools taught in the first year and used in the studio include:

- Rhinoceros and Grasshopper (www.rhino3d.com) for modeling
- Pollination/Ladybug Tools suite (www.pollination.cloud/), which incorporates Open Studio, EnergyPlus, Radiance and OpenFoam
- Design Builder as an alternate interface to Energyplus (https://designbuilder.co.uk/)
- Rhino CFD software is taught in parallel in Arch 7540 (www.food4rhino.com/en/app/rhinocfd)
- QGIS (www.qgis.org/en/site/)
- Climate Consultant 6.0 (www.sbse.org/resources/climate-consultant)
- Opaque 3.0 (www.sbse.org/resources/opaque)
- Window and Therm (https://windows.lbl.gov/software-tools)
- WUFI Hygro-Thermal modeling (https://wufi.de/en/)

Recommended Readings

DeKay, Mark, and G. Z. Brown. 2013. *Sun, wind, and light: architectural design strategies* (John Wiley & Sons).

Fathy, Hassan. 2010. *Architecture for the poor: an experiment in rural Egypt* (University of Chicago Press).

Geiger, Rudolf. 1965. *The climate near the ground* (Harvard University Press).

Givoni, Baruch. 1998. *Climate considerations in building and urban design* (John Wiley & Sons).

Heschong, Lisa. 1979. *Thermal delight in architecture* (MIT Press).

Moe, Kiel. 2010. *Thermally active surfaces in architecture* (Princeton Architectural Press).

Olgyay, Victor. 1963. *Design with climate: bioclimatic approach to architectural regionalism* (Princeton University Press).

Passe, Ulrike, and Francine Battaglia. 2015. *Designing spaces for natural ventilation: an architect's guide* (Routledge).

Student Work Examples

A New Chautauqua Institute (2018): The Weather Station, Houston

Students: Aishwarya Katta, Weston Huang, David Guarin
Studio Instructors: William W. Braham, Brian Phillips
Teaching Assistant: Mingbo Peng
Performance Design Workshop Instructor: Jihun Kim

The Weather Station team analyzed the thermal performance of an array of different configurations for the hot and humid climate of Houston. They discovered the importance of shading and ventilation for reducing the heating load. The first strategy was to raise the building and design an exterior screen wall to keep the sun off the building envelope, and then to add variably responsive openings to facilitate daylighting, views and ventilation (Figure 7.2).

As they developed the plan, they recognized that different uses were suited to different comfort regimes and that they could align those differences with the different temperatures that occurred in different parts of a free-running building. In the final layout, the core was devoted to program uses that required more stable temperatures, while more tolerant uses were situated in the corners. In the final bioclimatic comfort analysis, those corner zones were able to achieve comfortable conditions 75% to 82% of the time without any conditioning systems (Figure 7.3).

With steady refinement of the design, mostly reducing the cooling load with responsive shading and ventilation, but also reductions of the lighting and equipment loads, the team reduced the energy demand to work with the renewable power that could be harvested from the rooftop with combined solar photovoltaic and solar thermal panels (PV/T). A water-to-water heat pump system was used with a radiant cooling system to further enhance the COP of the system and reduce the amount of PV/T needed (Figure 7.4).

FIGURE 7.2 Responsive exterior façade. Weather Station, Houston.
Source: Aishwarya Katta, Weston Huang, David Guarin.

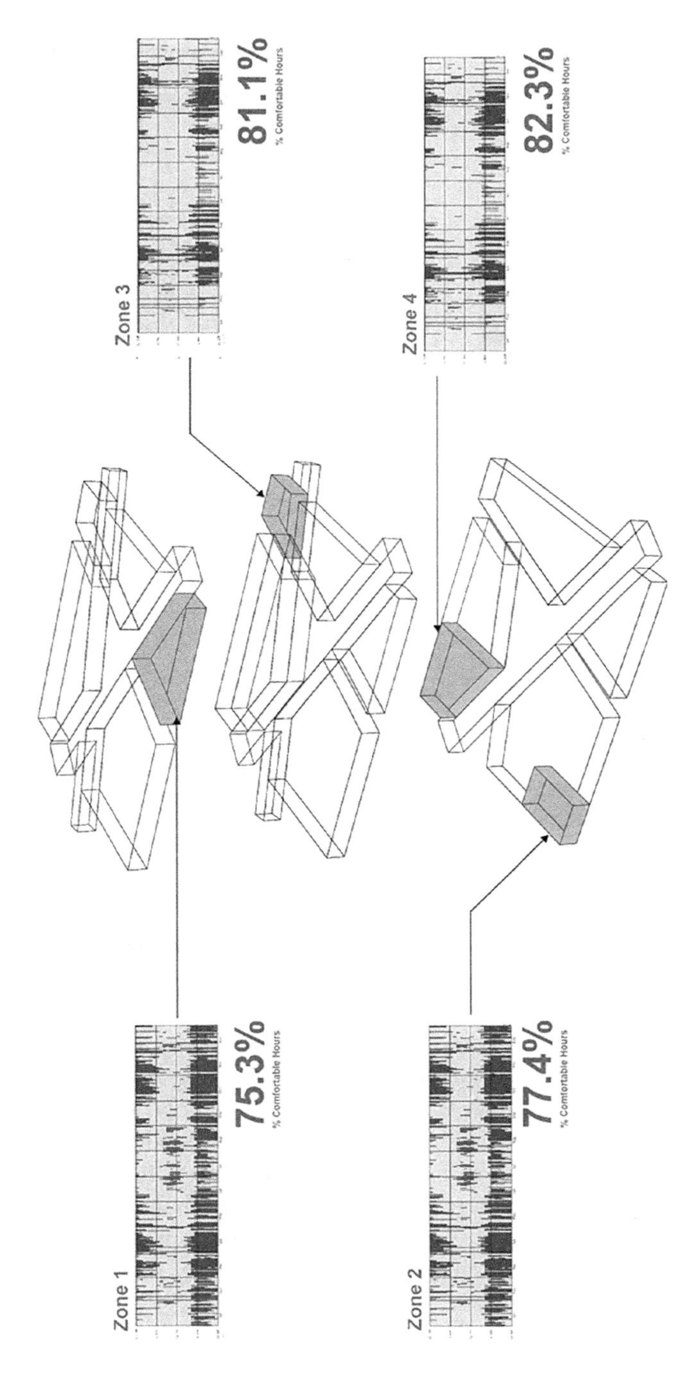

FIGURE 7.3 Adaptive comfort analysis of different zones and programs. Weather Station, Houston. Source: Aishwarya Katta, Weston Huang, David Guarin.

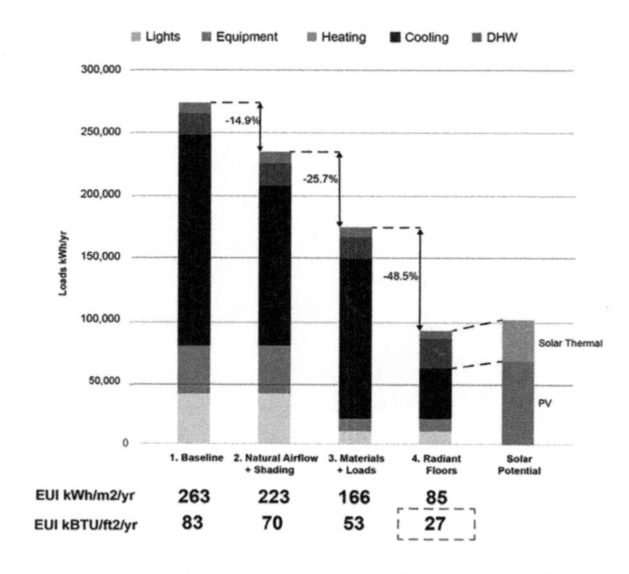

	Baseline	Final
Equipment Load W/m2	7.64	2
Lighting Density W/m2	11.84	3
Infiltration Load W/m2	0.0002	0.0001
Num of People per Area	0.0565	0.02
HVAC Type		
Cool COP	3.5	5
Heat COP	3.5	5.5
Air Flow Condition		
Min Indoor Temp (C)	x	20
Max Outdoor Temp (C)	x	28
Dorm Cooling (C)	x	>31
Dorm Heating (C)	x	<26
Class Cooling Pt (C)	x	>29
Class Heating Pt (C)	x	<26
Office Cooling Pt (C)	x	>29
Office Heating Pt (C)	x	<26
Cool Isld Cooling Pt (C)	x	>23
Cool Isld Heating Pt (C)	x	<16
Hpt Isld Cooling Pt (C)	x	>31
Hot Isld Heating Pt (C)	x	<26

Total Load: 864584 kWh/m2/yr

Solar PV Potential: (60% roof space/ panel area sqft) x 22% efficiency

Solar Thermal Potential 40% roof space x 25 W/ft2 x 204 days x 12 hours

FIGURE 7.4 Performance analysis of iterations of project design. Weather Station, Houston.
Source: Aishwarya Katta, Weston Huang, David Guarin.

Reimagining the Urban Data Center (2020): The Wind Cube, Seattle

Students: Junjie Lu, Yaning Yuan
Studio Instructor: Dorit Aviv
Teaching Assistant: Zherui Wang
Technical Assistant: Kit Elsworth
Performance Design Workshop Instructor: Jihun Kim

This design for a data center located in downtown Seattle leveraged Seattle's moderate climate and facilitated direct use of outdoor air to cool the internal load produced by the servers. The building form was derived from analysis of airflow and waste heat from servers. The design aims to simultaneously satisfy the cooling needs of server spaces, the heating needs of the adjacent office spaces, and to optimize natural ventilation airflow. Both wind and buoyancy forces were used as guidelines to determine the shape of the internal openings and air paths, which in turn define the overall shape of the building. Using CFD simulations, streamlines became air channels in the building mass to allow for effective circulation of cool air. Furthermore, a heat distribution system was incorporated into the design for the recirculation of waste heat into a rooftop greenhouse.

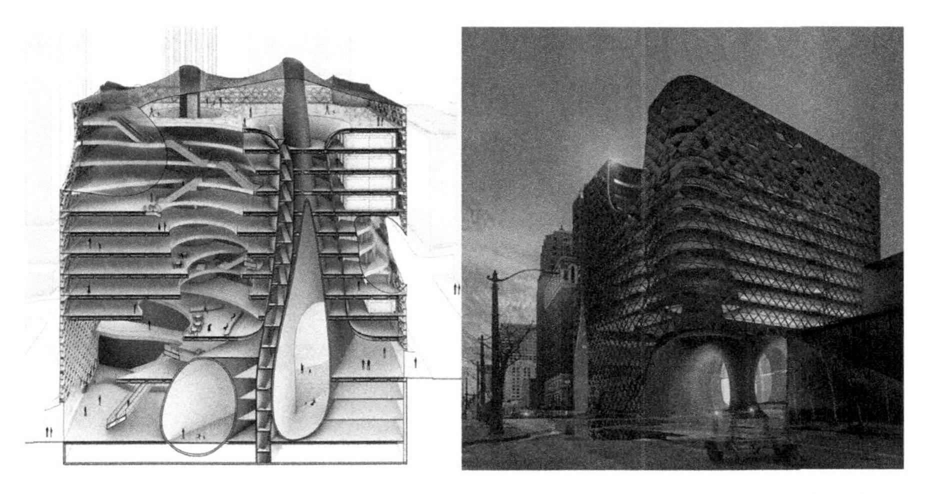

FIGURE 7.5 Section and rendered view of Wind Cube, data center and office building. Wind Cube, Seattle.
Source: Junjie Lu, Yaning Yuan

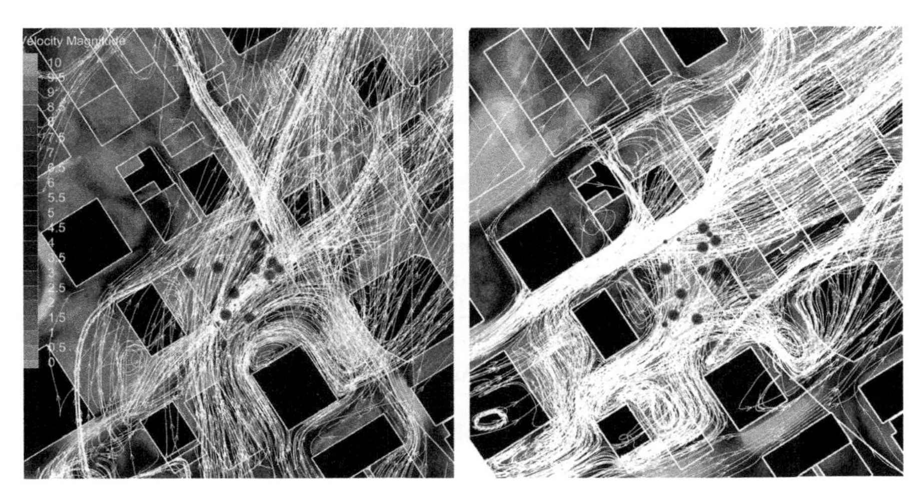

FIGURE 7.6 Urban wind flow analysis. Wind Cube, Seattle.
Source: Junjie Lu, Yaning Yuan

Naturally Ventilated Schools (2021): The Wing School, Singapore

Students: Jun Xiao, Qi Yan
Studio Instructor: Dorit Aviv
Teaching Assistant: Suryakiran Jathan Prabhakaran
Performance Design Workshop Instructor: Jihun Kim

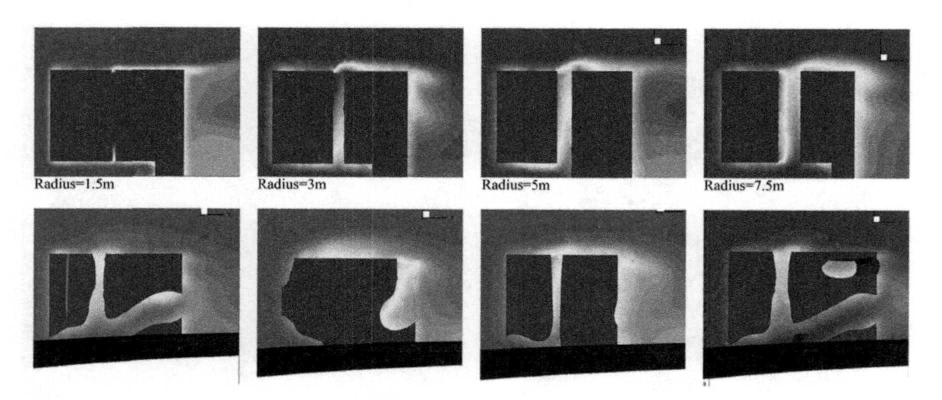

FIGURE 7.7 CFD analysis of air flow in building section. Wind Cube, Seattle. Source: Junjie Lu, Yaning Yuan

The Wing School was sited in Singapore, whose climate is always hot and humid, with few environmental heat sinks available. The design creates a naturally ventilated and shaded space for children, allowing them to experience air drawn in from the surrounding gardens. With close contact to the outdoor environment – wind, rain and sunlight – children would develop a sensibility towards nature. The design closely followed an urban wind simulation to select the most appropriate wind pattern from more than 30 layouts, and refined it based on the urban context of the site. The design was developed into a curvilinear form with two publicly accessible playgrounds. Considering the need for shading and shelter from the rain, it was configured with a large roof and several recessed openings which act as rain collectors. The layout of each floor was refined for daylighting and shading patterns. Large, tree-like wooden structural columns were integrated within the roof structure and incorporated vertical circulation and enhancing air flow. The team also added radiant cooling pipes supported by the branching structural columns to ensure thermal comfort while relying mostly on natural ventilation in the hot-humid climate.

Outreach Opportunities

The work is published on the program website (https://ebd.design.upenn.edu/), and in *Pressing Matters*, the annual publication of the Department of Architecture.

Course Assessment

The student work is assessed by expert reviewers in the middle and end of the semester, including architects, engineers, and building scientists. Students present and defend design proposals and the results of their performance analysis. The studio is also paralleled by *Performance Design Workshop*, which helps students develop research techniques to evaluate their design proposals. That course

FIGURE 7.8 Views and temperature gradient simulations of naturally ventilated schools. The Wing School, Singapore.
Source: Jun Xiao, Qi Yan

provides the opportunity to iteratively review and assess their analytical work so that claims about thermal, energy, and carbon performance are well validated.

The graduates of the programs find positions in the specialized sustainability or research units in architectural practices and also in consulting practices, especially those that specialize in early design.

Looking Ahead

While the basic approach of the Bioclimatic Design Studio remains the same, it continues to evolve, taking on new programs and building types that are most relevant to investigate as the world changes, and using new tools and methods to evaluate the local microclimate and environmental resources such as evaporation, radiative sky cooling, and ground exchange.

Note

1 Victor Olgyay, *Design with Climate: Bioclimatic Approach to Architectural Regionalism*, Princeton, NJ: Princeton University Press, 1963.

8

A PHYSICAL AMBIENCES APPROACH TO THE ASSESSMENT AND REPRESENTATION OF CARBON NEUTRAL ARCHITECTURE

Claude M. H. Demers and André Potvin

Course Overview

Course Names

Design-Research Track on Physical Ambiences (2 courses)

- Physical Ambiences and Design Studio (ARC-6037)
- Architectural and Urban Physical Ambiences Seminar (ARC-6044)

University / Location

Université Laval / Quebec City, Québec, Canada

Targeted Student Level

- Graduate
  - 1st year, Master of Architecture, 2-year program with optional 1.5-year advanced standing

Course Learning Objectives

In this course, students will:

- Understand the fundamental embedded concepts of physical ambiences at the urban, architectural and material scales.
- Respond to the physical characteristics (thermal, luminous and acoustical) of a site in developing a strategy for controlling ambiences in a project.

DOI: 10.4324/9781032692562-11

- Analyze and predict ambiences qualitatively and quantitatively using analogue and/or digital tools.
- Develop student's ability to control physical environments through architectural variables first to reach carbon neutrality.

Course Learning Outcomes

Successful completion of this course will enable students to:

- Generate an integrated response to complex societal challenges through responsive carbon neutral architecture.
- Represent the qualitative and quantitative dimensions of physical ambiences.
- Develop a common interdisciplinary language essential for a genuine integrated design process.

Course Methodology

Philosophy

The Physical Ambiences approach proposes a poetic and fundamental reflection on the roots of the environmental movement through the investigation of matter and energy fluxes at all scales of the architectural design process, from inception to detailing. Among perceptible qualities of the built environment, three interdependent physical phenomena particularly define nature-architecture transactions: the thermal, lighting and acoustical dimensions. Their systemic study within the context of bioclimatic design brings forward the notion of sustainable genius loci stemming primarily from the variables of architecture. Physical ambiences tackle the complex equation between quantitative (energy, matter) and qualitative aspects (perception of comfort, wellbeing and aesthetics) of the built environment. Figure 8.1 illustrates relations between key concepts discussed in the pedagogy of the Physical Ambiences approach. It builds on the sensitive universe of future

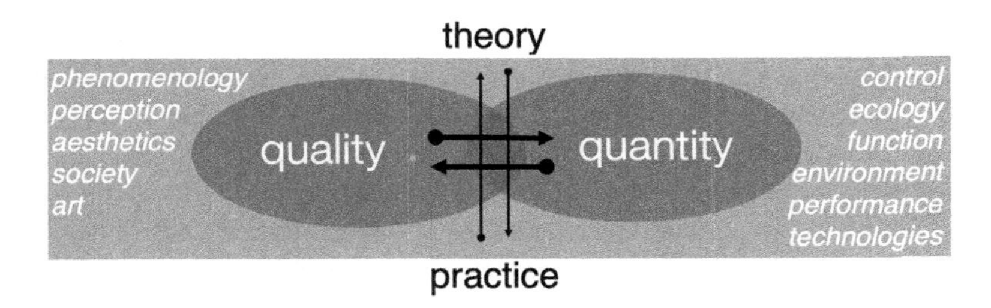

FIGURE 8.1 The systemic nature of physical ambiences.

architects upstream of the project, improving the capacity to deliver ideas within a coherent system of material and energy fluxes responding to inhabitants' needs and their ecological ambitions. The assessment and representation of physical ambiences therefore occurs at the intersection of theoretical/conceptual and applied knowledge.

In the current race towards carbon neutral/net-zero design where numerical optimization tools often command, the physical ambiences approach reiterates the importance of analogical "hands-on" experimentations through physical models as a critical thinking foundation for an embodied sustainable design intention.

Definitions

Carbon neutrality – Net-zero carbon emissions over the entire lifecycle of a building including human agency where inhabitants' behavior becomes the main determinant of performance.

Environmental performance – Frugality first, complexity when needed. Acknowledgement of the importance of robust and resilient design strategies for the generation of passive low energy architecture.

Physical ambiences – Systemic combination of thermal, lighting and acoustical environmental factors through architectural variables and the integration of environmental control systems.

Implementation

The Physical Ambiences approach involves graduate students enrolled in a design research concentration of the professional Master of Architecture, which combines a graduate design studio and its concomitant seminar for an entire semester. The Physical Ambiences concentration was created and jointly taught by Demers (daylighting) and Potvin (thermal comfort) since 2001. Both professors have developed analogical tools and calculation methods to assess ambiences through existing environments[1] and models.[2]

Demers and Potvin are equally involved in each aspect of the course, ensuring an integration of all types of knowledge, from the qualitative aspects of the human experience of ambiences to the measurable quantitative metrics of physical environments. Over the years, the team has developed fundamental research outcomes applicable to the Physical Ambiences design process within the Groupe de Recherche en ambiances physiques (GRAP).

Interdisciplinary collaborators from other departments are especially involved in the design studio to address technological integration, including mechanical engineering, structure and wood engineering (Figure 8.2). An acoustical specialist acts as a regular collaborator at the School of Architecture. Community partners and professional architects are invited guests at all phases of the design

FIGURE 8.2 Interdisciplinary collaborations through academic, community and research partners.

studio project. Research partners may take part in special thematic studio days, for instance to tackle simulation results from software analysis, or to integrate an emergent subject or expertise.

Assignments

The Physical Ambiences design studio combines theoretical and design research results in the generation of architectural hypotheses, which are developed with environmental assessment tools and rules of thumb. The objective of the studio is to interpret and integrate data analysis and results to generate an informed evidence-based design throughout the semester. The seminar is the companion course that combines fundamental and empirical research-based knowledge to inform the design studio activity. It aims at the exploration of analysis tools, methods, and metrics in the generation of relevant data results at each design phase. Figure 8.3 illustrates the "theory into practice" schematic framework of

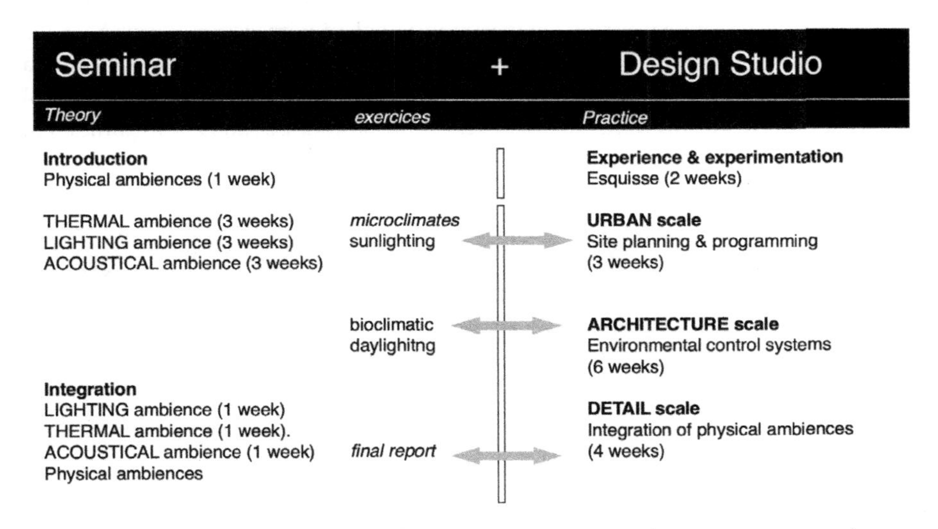

FIGURE 8.3 Schematic framework of combined Physical Ambiences courses and content structure.

the combined courses. It highlights the relationships between seminar contents and design studio phases during the academic semester. Exercises of the seminar are coordinated with design studio crits to ensure migration of contents directly into architectural solutions throughout the project.

A conceptual liberating *esquisse*, consisting of a preliminary design exploration, serves as a catalyst for stimulating imagination and guiding subsequent studio endeavors focused on the intersection of environmental performance and users' embodied experiences of spaces. This initial exercise systematically tackles various aspects related to physical ambiences, including the visualization of human experiences, the exploration of tools and techniques for experimenting and envisioning architecture, and the recognition of the importance of simple representation skills, such as drawings and images, to effectively synthesize complex sets of results. Figure 8.4 shows the conceptual development of a structural grammar based on folding structures, from experimentation to experience. The organized set of models is studied under sun lighting (Figure 8.4. left), illustrating the effects of shape and shadows on the conceptualization of intermediate spaces. Photographic explorations under the real sky (Figure 8.4 right) aim to translate conceptual ideas into inhabitants' experiences, linking structural explorations with space appropriation. The holistic approach of this preliminary stage lays the foundation for a nuanced understanding of the intricate relationship between design, environmental factors and the human experience of place.

Design outcomes are the result of a reflection based on theory, research and practice into the pedagogy of the architectural project. Assessments of environmental factors such as daylighting, natural ventilation and thermal heat losses and solar gains are key components of bioclimatic architecture, based on empirical knowledge gained through iterative actions and intuitive learning activities. Thermal and daylighting assignments consist of seminar reports linked with design projects, at the urban, architectural and detail scales. In these exercises, students learn to identify a path to carbon neutrality by formulating a design-research question. Architectural precedents help identify variables that become the basis of preliminary hypotheses to be further experimented and analyzed with physical and digital models.

FIGURE 8.4 Experience and experimentation: an exploratory *esquisse* as an introductory design exercise.

Source: R. Boulet, A. S. Boutin and X. Sgobba, Université Laval; Supervisors: C. Demers and A. Potvin

The art of communicating innovative carbon neutral design strategies is an integral part of the design studio learning outcomes. Exploration of several media representations allows for the illustration of both qualitative and quantitative data throughout the design process. Participating in architectural design competitions also becomes a powerful driver for synthetic graphical representations that capture the essence of the architectural propositions where carbon neutrality and environmental performance coincide with design excellence. This pedagogical approach underscores the practical integration of theoretical principles into the applied context, fostering a comprehensive understanding of the symbiotic relationship between design aesthetics and sustainable practices.

Analytics

Digital simulation tools are rapidly evolving, with varied limitations in terms of results and output. Throughout the years, several software was tested to find the most exhaustive one that could be used at all design stages. However, when it comes to critical thinking, students must be given several options to test and compare different types of simulations, from analogical physical modeling under real skies or in the artificial sky, to digital simulations. GRAP developed custom design tools used in the early design phases to progressively build student literacy and confidence in their skills. The assessment of the three ambiences therefore begins with simple analytical spreadsheets and nomograms, providing design hypotheses that are later validated with physical modeling and ultimately digital simulations. Figure 8.5 shows the main tools that relate to each ambience and laboratory in the Physical Ambience approach. Tools become part of the design activity, closely linking analytic validation and spatial representation.

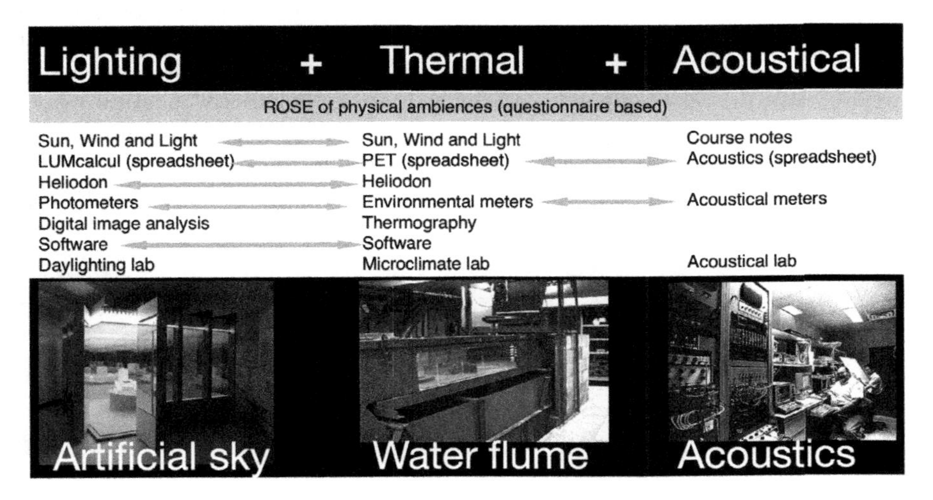

FIGURE 8.5 Integration of numerical and analogical tools for lighting, thermal and acoustical ambiences.

Visual ambiences assessment begins with LUMCalcul, a custom spreadsheet by Demers[3] that emphasizes on general daylighting indicators that are later validated with physical models under diffuse and direct sky using heliodon and a walk-in artificial sky. The advantage of LUMcalcul is that it adopts a more intuitive relationship with design sketches and representation tools such as architectural section drawings. Photocell measurement provides quantitative results, while photography offers digital pattern analysis for qualitative analysis. Figure 8.6 shows a series of design activities aiming to validate the generation of a space in relation to daylighting. The structural model study (Figure 8.6, left) enables testing of several building envelopes. The physical model creates opportunities for color filters analysis (Figure 8.6, middle image, top), glare analysis using high dynamic range imaging (Figure 8.6, middle image, bottom), to perspective renderings including inhabitation possibilities of the space, such as in Figure 8.6, right. More complex indicators such as daylight autonomy are assessed with specialized digital tools.

Thermal ambiences assessment also begins with PET (Profils d'équilibre thermique), a simple custom spreadsheet by Potvin[4] providing thermal balance profiles according to program, climate and envelope that help identify the most critical passive strategies determining thermal performance. Further digital simulations optimize comfort provisioning using the thermal adaptive model. Microclimatic conditions at the urban/site scale are assessed using a waterflume to optimize winter and summer outdoor comfort. Figure 8.7 illustrates the resulting bioclimatic section drawing (top of image) validated from natural ventilation studies using tracing ink in the waterflume (bottom sequences of images). Such methodology contributes to the refinement of indoor-outdoor spaces, aligning them with optimal comfort parameters to develop innovative design proposals and rigorous scientific inquiry.

By integrating all aspects of physical ambiences early in the design process, simple analytic tools and physical models accelerate the daylighting and thermal experiential skills that link to other physical ambiences through rapid haptic exploration (the thinking hand). It helps validate quantitatively (physical model measurement) and qualitatively (perception) the design intentions with several collaborators at the studio; and connects all types of data in a single representation through digital overlay and sketching. Numerical models complement physical

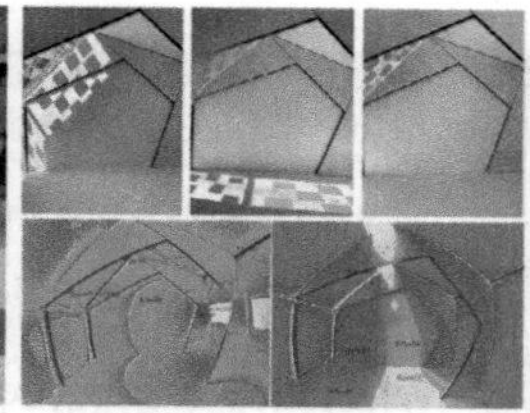

FIGURE 8.6 Daylighting and space generation.
Source: A. Cliche, I. Isabel and T. Rouleau-Dick, Université Laval; Supervisors: C. Demers and A. Potvin

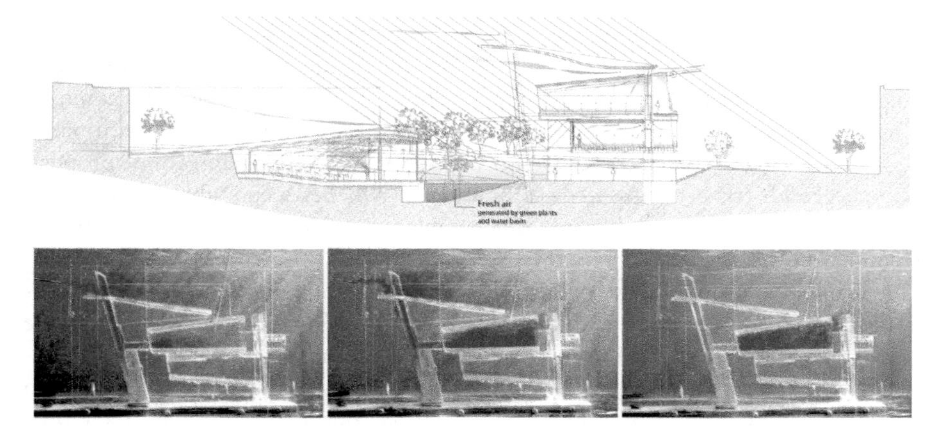

FIGURE 8.7 Bioclimatic representation with corresponding analogical waterflume natural ventilation performance simulations.
Source: E. Boudreault-Sauvageau and P. Gobeil, MArch, Université Laval; Supervisors: C. Demers and A. Potvin

models and validate results according to accepted visual and thermal metrics. The selection of simulation software varies according to design requirements with a special emphasis on their potential for data visualization. The comprehensive Physical Ambiences approach ensures a synergistic integration of various elements, fostering a holistic understanding of the design process.

Required Reading

DeKay, Mark and G. Z. Brown. *Sun, Wind and Light: Architectural Design Strategies*. New York: John Wiley and Sons, 3rd edition, 2014. The book provides general guidance of bioclimatic strategies to apply preliminary dimensioning of architectural variables.

Recommended Reading

Adolphe, Luc. *Ambiances Architecturales et Urbaines*. Cahiers de la recherche architecturale, n° 42–43, Éditions Parenthèses, 1998. The document consists of a selection of texts referring to holistic approaches to ambiences as human experience in architecture.

Other references vary according to specific objectives of a studio context, often selected from research articles on qualitative and quantitative measure of ambiences produced by GRAP graduate students. Demers and Potvin published on the importance of an analogical approach to architecture based on studio work[5] and design research.[6]

Outreach Opportunities

Collaborative design research projects with communities allow a grounded reflection based on inhabitants' sustainable needs and aspirations. Recent editions of the studio integrate a reflection on the UN's Sustainable Design Goals to cast a more global perspective on the architect's contribution towards carbon neutrality. The design studio allows a discussion on the common understanding of proposed metrics for the assessment of specific performance goals, offering a vivid perspective on the acceptability of architectural propositions. Field trips to isolated communities, whenever possible, provide an opportunity to further integrate local knowledge and validation with and by the community of the design hypotheses.

The integration of external examiners and community partners is instrumental in fostering reflexive thinking and sound judgment, achieved through the active engagement of end-users, decision-makers and professionals in the evaluation process. Professional architects participate to and in-between crits to validate the feasibility of students' propositions but also to emulate their own practices through the fresh and disruptive qualities of academic work. Furthermore, interdisciplinary collaboration with structural engineering, wood engineering, mechanical engineering and agriculture proves pivotal in cultivating integrative design approaches and fostering a shared understanding of sustainability and carbon neutral design. Several students of this MArch track on Physical Ambiences have extended their interest into formal MSc, or PhD research-design thesis.[7] As teaching assistants, they close the loop by further developing tools and techniques that have been validated in the design studio. The participation of graduate students also adds a dynamic dimension by bridging the gap between academia and practitioners. This collaborative effort not only contributes to the advancement of sustainable design but also serves as a catalyst for pedagogical and practical innovation within the field of sustainable design.

Course Assessment

The pre-professional studio-seminar course concentration on Physical Ambiences addresses a wide array of the Canadian Architectural Certification Board (CACB) accreditation criteria, including Comprehensive Design, with an emphasis on Technical Knowledge related to Envelope systems and Environmental systems. These aspects are often seen as too technical to contribute to the more theoretically oriented architecture studio culture. The Physical Ambiences approach therefore seeks to transcend the existing dichotomy between technical expertise and design theory. Results aim to provide valuable insights to the ongoing discourse surrounding the interdisciplinary nature of architectural practice, with the ultimate goal of fostering a more holistic and collaborative approach to design within architectural studios.

The interpretation of data and metrics from various tools and simulation techniques may initially appear overwhelming to students, but essential to develop a critical perspective on data results and analysis. In the realm of environmental performance, it is imperative to address fundamental aspects and provide simplified representations that students can discuss in a critical way. The selection of the relevant information plays a pivotal role in validating innovative design solutions, necessitating a tailored approach and guidance aligned with specific project goals. Lessons learned encompass building on students' interest in both analogical tools and techniques, supplemented by numerical tools. The incorporation of these tools is essential, and their integration should be executed in a clear manner, emphasizing visualization and representation that hold relevance for architects and potential users alike.

Student awards:

- 2019 to 2024 AIA COTE Top Ten for Students Competition (8 winning projects)
- 2016 Blue Award, wood innovation (mention)
- 2014 Autodesk AIA Sustainability (2nd prize, mention)
- 2012 ACSA Architecture and Engineering of Sustainable Buildings (1st prize)
- 2011 International Design Competition "Ajout manifeste" (1st prize)
- 2007 ACSA Concrete Design (1st, 2nd prizes, mention)
- 2004 ACSA Laboratory for the 21st century (2nd prize, mention)
- 2003 ACSA Wood Design Council (mention)

Looking Ahead

Reflecting on a two-decade pedagogical experience with the Physical Ambience approach, several key conclusions emerge. First and foremost is the recognition of the significance of revisiting the fundamental roots of environmental design research and theory, encapsulated in the principle of *looking back to the future*. Emphasizing the role of design-research methodology as a pivotal driver for empirical knowledge acquisition, the conclusion underscores the importance of experiential or "acquired" knowledge. Visualization of physical ambiences at all design stages through imagery and metrics is the key idea that can objectively translate into clear and communicable terms. Moreover, the synthesis of architectural design and research is highlighted as a powerful source of meaningful hypotheses, requiring validation through both fundamental and practical research experimentation. In this context, architecture emerges as the focal point of the integrated design process, embodying a central role in the interplay between theory and application.

Notes

1 A. Potvin, "Movement in the Architecture of the City: A Study in Environmental Diversity," PhD thesis, The Martin Centre for Architectural and Urban Studies, University of Cambridge, 1997, DOI: 10.17863/CAM.31085.

2 C. M. H. Demers, "The Sanctuary of Art: Images in the Assessment and Design of Light in Architecture," PhD thesis, The Martin Centre for Architectural and Urban Studies, University of Cambridge, 1997, DOI: 10.17863/CAM.16320.

3 C. M. H. Demers and André Potvin, "On the Art of Daylighting Calculations: LUMcalcul as a Prediction Tool in the Early Design Stage," *PLEA* (2013): 510–547.
4 A. Potvin, C. Demers and H. Boivin. "PETv4. 2 Les profils d'équilibre thermique comme outil d'aide à la conception architecturale," *Proceedings of eSIM2004, Vancouver* (2004): 9–11.
5 Claude M. H. Demers and André Potvin, "Erosion in Architecture: A Tactile Design Process Fostering Biophilia," *Architectural Science Review* 60, no. 4 (2017): 325–342, DOI: 10.1080/00038628.2017.1336982.
6 C. M. H. Demers and A. Potvin, "From History to Architectural Imagination: A Physical Ambiences Laboratory to Interpret Past Sensory Experiences and Speculate on Future Spaces," *Ambiances* 2 (2016): http://journals.openedition.org/ambiances/756, DOI: 10.4000/ambiances.756.
7 M. Watchman, C. Demers and A. Potvin, "Towards a Biophilic Experience Representation Tool (BERT) for Architectural Walkthroughs: A Pilot Study in Two Canadian Primary Schools," *Intelligent Buildings International* 14, no. 4 (2022): 455–472, DOI: 10.1080/17508975.2021.1925209; M. Watchman, M. Mark DeKay, C. Demers and A. Potvin, "Design Vocabulary and Schemas for Biophilic Experiences in Cold Climate Schools," *Architectural Science Review* 65, no. 2 (2022): 101–119, DOI: 10.1080/00038628.2021.1927666; P. Charest, A. Potvin and C. M. H. Demers, "Aquilomorphism: Materializing Wind in Architecture through Ice Weathering Simulations," *Architectural Science Review* 62, no. 2 (2019): 182–192. DOI: 10.1080/00038628.2018.1535423; L. Mazauric, C. M. H. Demers and A. Potvin, "Climate Form Finding for Architectural Inhabitability," *Ambiances* 4 (2018), accessed December 5, 2018, http://journals.openedition.org/ambiances/1688. DOI: 10.4000/ambiances.1688.

9

ENVIRONMENTAL BUILDING DESIGN RESEARCH STUDIO

Design Innovation for the Climate Emergency

William W. Braham and Billie Faircloth

Course Overview

Course Name

Arch 7090: Environmental Building Design (EBD) Research Studio with Arch 7550: Environmental Innovation and Prototyping

University / Location

University of Pennsylvania / Philadelphia, PA, USA

Targeted Student Level

- Graduate
    - 2nd year, 3rd semester, post-professional Master of Science in Design with a concentration in Environmental Building Design (MSD-EBD)

Course Learning Objectives

In this course, students will:

- Understand the impacts of the climate emergency on people and communities and the architect's role in this crisis.
- Classify climate adaptation structurally, physically and biologically, and develop critical arguments on design for adaptation discourse.
- Prioritize building reuse (2022) or bioclimatic design (2023) as a climate adaptation strategy of cultural and ecological significance.

DOI: 10.4324/9781032692562-12

Course Learning Outcomes

Successful completion of this course will enable students to:

- Develop a design research approach informed by systematically collecting, organizing, querying and interpreting data for comfort, energy, carbon, health and equity.
- Create an architectural intervention using a testable, measurable design hypothesis and appropriate architectural computation methods.
- Develop in-depth knowledge of design for climate adaptation by detailing and delineating a project through analysis, drawing, modeling, simulation and physical prototyping.

Course Methodology

Philosophy

The EBD Research Studio is a required course in the third semester of the post-professional degree program, Master of Science in Design in Environmental Building Design (MSD-EBD). The program's first semester features coursework on building science, building performance simulation, and the theory and practice of environmental accounting, which is then synthesized in the second semester through a Bioclimatic Design Studio, a Performance Design Workshop, and course in the History and Theory of Architecture and Climate. In the program's third semester, an independent course on Innovation and Prototyping parallels and supports the objectives of the Research Studio.

The EBD Research Studio is dedicated to design innovation, which involves formulating new approaches to environmental building design and testing and validating those approaches. Student teams, building upon their previous course work, are challenged to develop research proposals in alignment with the topic of the year and criteria for evaluating its outcomes. This chapter uses examples from two years in the studio, both framed as bioclimatic adaptations to the climate emergency. The 2022 studio focused on adapting an historic high-rise dormitory to the conditions of climate change, while the 2023 studio focused on adapting to conditions of extreme heat.

In the original formulation developed by Olgyay, bioclimatic design involves four steps: analyze the climate, evaluate its effect on human comfort, calculate the effect of technological interventions, and synthesize those in an architectural proposal.[1] Bioclimatic adaptations must balance human/community-centered, technical and ecological concerns. Students use a design research process, integrating inductive and deductive methods, to form a hypothesis, a study question and a research proposal. Using the design process, students implement their research proposal through successive rounds of speculation and analysis to discover a deeply informed and validated design proposal.

Definitions

Carbon neutrality – Not a central term in the studio's work, though students concluded that climate change adaptation is a process of mitigating risk for uncertain future conditions and demands multiple forms of adaptation.

The studio's terminology included the following terms:

Active systems – Systems which use extracted fuels or resources to power conditioning systems.

Adaptive capacity – The ability of a building and its occupants to deal with future climate disruptions.

Circularity – The recycling or reuse of materials in a Circular Economy, including accounting for the costs of recycling/reuse.

Climate adaptation – Intentional changes in buildings, infrastructures and social arrangements to accommodate climate change.

Degrowth – A movement which argues that growth is itself a cause of environmental destruction and which supports the ideas of steady-state or circular economies.

Living laboratory – A place-based research unit that serves as a test bed for innovative practices. It typically combines technical and social innovations, with real-time feedback and evaluation.

Mitigation – Reducing the emission of greenhouse gases through greater efficiencies or switching to a low-carbon power supply.

Passive survivability – The capacity of a building to remain inhabitable during disruptions to supplies of power, fuels and water, though some studio teams extended it to include food.

Passive systems – Systems which use environmental energies or resources to power conditioning systems. We included both purely passive systems and those that use power for fans, pumps, or vapor compression cycle devices to utilize locally available resources or heat sources and sinks.

Resilience – The ability of a building or system to recover from a disturbance and usually requires some additional resources or capacity.

Implementation

2022 EBD Research Studio

The 2022 studio began with a study question: *How can designers challenge attitudes towards structures, systems and behaviors to demonstrate an approach to climate adaptation through building reuse?* Buildings contribute significantly to planet-killing emissions, especially those in developed countries and cities. If, according to Carl Elefante, "the greenest building is the one that is already built," then focusing on adaptive reuse, and renovation could be one of the most important forms of architectural disruption.[2] This is especially true when we recognize that adaptation occurs across varied spatial and temporal scales involving our bodies, things, whole buildings and migration.

The studio explored approaches to the climate emergency by developing bioclimatic adaptations for people and their habitats through a *living laboratory* mindset emphasizing the importance of continuously learning about adaptation using built environment interventions. Strategies were developed and tested using Sansom Place West, a University of Pennsylvania residence hall designed by Richard and Dion Neutra and completed in 1970. After more than 50 years of use, all systems in this 16-story, double-loaded corridor building required rethinking and retrofit. Over 16 weeks, student teams developed proposals for bioclimatic interventions, modifying the building's program, purpose and systems, considering people's accessories and practices.

Student teams discovered that design for climate change adaptation is a process of mitigating risk for uncertain future conditions and demands multiple forms of adaptation. Through their research, 5 student teams each identified a primary strategy to achieve reductions in carbon emissions and adaptation to the future. As one of the student teams observed, "The discourse of climate adaptation has challenged our studio to interact with time as an element of the work."

2023 EBD Research Studio

The 2023 studio explored architectural approaches to extreme HEAT, developing bioclimatic adaptations for people in tropical climates experiencing heat extremes now exacerbated by global warming. Bioclimatic strategies drew on all available environmental resources in design proposals for a desert shelter whose performance was tested with digital and physical prototypes.

Student teams explored all sides of the bioclimatic proposition—the ways that buildings interact with and create microclimates, the adaptability of people to climatic conditions, and the many techniques for mediating between the two. Traditional forms of bioclimatic design have largely employed passive methods such as shading, insulation and mass construction, but these reach a limit in extreme HEAT, and require additional techniques to utilize environmental resources and heat sinks. Evaporation, radiant exchange with the sky, and deep ground temperatures all provide pathways for rejecting heat, while careful use of buffer zones, gradations of shade, air movement, radiant cooling and dehumidification allow buildings to adapt more flexibly.

Assignments 2022

The 2022 studio approached its design research through three overlapping "layers."

Layer 1 // Defining

What are the factors that contribute to the act and process of adapting?

Students engaged mind mapping, using their prior knowledge to discover structural, physical, atmospheric and biological forms of adaptation. They identified multiple conditions of change, including climatic factors, migration, and social and economic reorganization. The task of *defining* continued through a topic-specific literature review. Each topic related to adaptation, from ones that were specific to buildings, such as envelopes, materials, and active and passive systems, to others that were more general, such as growth, resilience, mobility/migration and ecology. Each topic was distilled into critical study questions that yielded useful outcomes and strategies that could inform design. The result was consolidated in a *Climate Adaptation Playbook* shared within the studio.

As one of the student teams concluded, "Over the course of our program, we have become acutely aware of energy flows through buildings, components, and building systems ... [and recognized] that increased building performance alone will not solve the climate crisis."

Layer 2 // Dissecting

What is the form, function and fitness of Sansom West and its systems?

Student teams studied Sansom West to develop a comprehensive understanding of its site, systems, spatial distribution, assemblies and future climate. The studio modeled its systems and simulated the building's thermal, energy and daylighting behavior, performed targeted analysis across selected environmental and material flow criteria and used metered and sensor data to augment its understanding of building and inhabitant behavior. The task of *dissecting* established the basis for constructing a design for the climate adaptation hypothesis and research proposal.

Layer 3 // Adapting

What is an adaptation-first approach to Sansom West's reuse, and who does it benefit and how?

In the final phase of research, student teams identified stakeholder groups, primary design topics, and adaptation strategies to both the changing climate and the conditions changed by the changing climate. The strategies of adaptation guided the development of comprehensive projects for a *living lab* at Sansom West, requiring students to consider the social, cultural and financial aspects of the adaptation. The design process cycled between deeper topical research, design exploration and performance analysis with analytical models, digital simulation and physical prototypes as appropriate to their topic.

Assignments 2023

The 2023 studio approached its design research through four overlapping "layers."

Layer 1 // Catalogue

Studio teams examined strategies of adaptation to conditions of extreme HEAT. They explored structural, physical and biological adaptations, collecting instances into a catalogue for use by the studio. These formed the basis of research agendas for studio teams to test in Hot Box experiments and further elaborate in designs for a desert shelter. The outcome was a ***Heat Catalogue*** that expanded the studio's vocabulary and deepened its understanding and grasp of HEAT's spatial and temporal effects.

Layer 2 // Hot Box

Studio teams tested initial strategies for keeping interior environments cooler under conditions of extreme HEAT, using a heated climate chamber (hot box) with a strong radiant source to conduct experiments on the movement of heat in different materials and configurations. The Hot Box was kept at a constant, high temperature (~35C) and the radiant source was turned on and off in a 15-minute cycle. Three phases of testing explored (1) configurations with the same insulating board, (2) configurations with different materials, and (3) configurations with the addition of water.

Layer 3 // Desert Shelter

Based on the research of the first 2 layers, each team developed a design and testing proposal for a shelter in the desert near Taliesin West, where the Taliesin fellows built their winter shelters. The studio's buildings provide year-round housing for residents involved in research on the architecture of extreme heat. Each desert shelter housed 2 people with basic kitchen and bath facilities. Teams selected specific sites in the desert of roughly 6 m by 6 m, with up to 3 m of the surrounding area serving as landscape or buffer space.

Layer 4 // Refining, Prototyping, Testing

In conjunction with the course on *Innovation and Prototyping* (7550) each team identified a bioclimatic technique to explore and articulated a research question to answer. They prepared a literature search on the technique and identified the opportunities for innovation, the methods of analysis and metrics for evaluating the results. They prepared a plan for prototyping the bioclimatic technique, including digital analysis and physical prototyping,

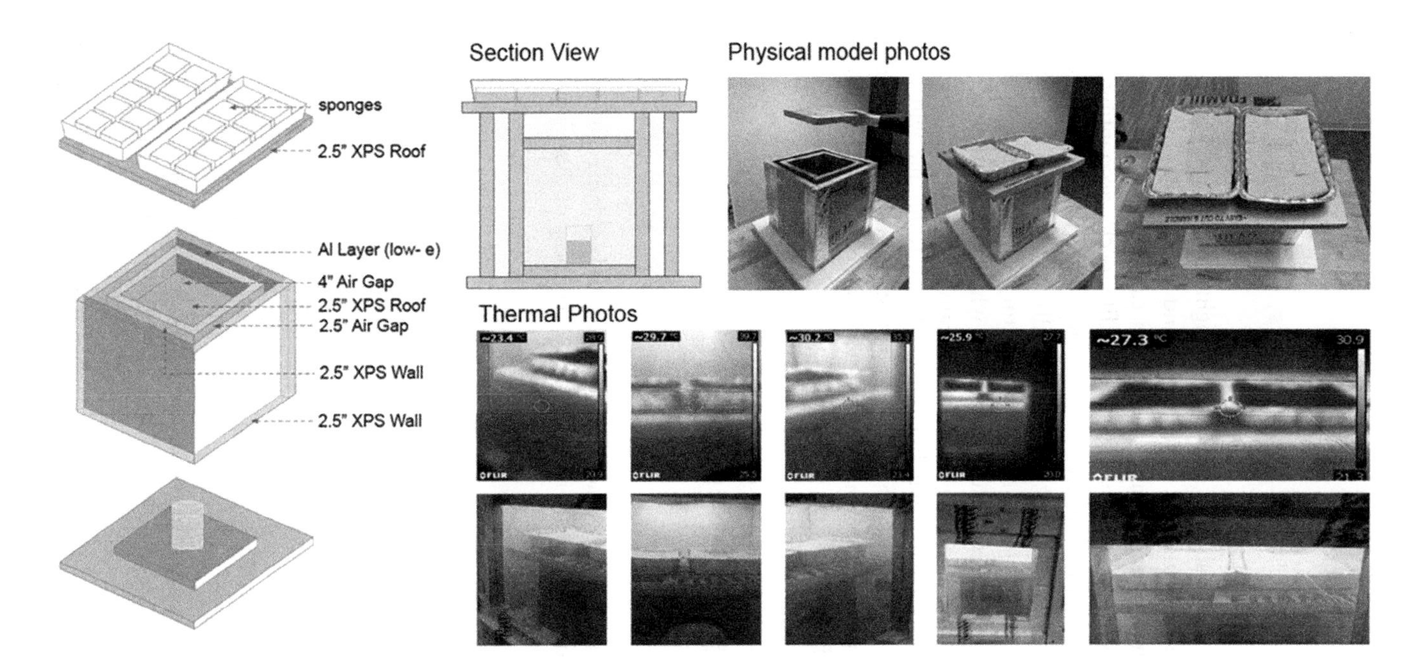

FIGURE 9.1 Hot box experiment, phase 3, showing final configuration with evaporative panels on the roof.
Source: Dishanka Kannan, Ningxu "Janvier" Luo, Yujia Cui

including methods of measurement, materials and schedule of testing. The results of the testing were used to support the studio and are being published as conference or journal papers.

Analytics

The studios draw on the analytical techniques taught in the first year of the program, including comfort analysis, heat, energy and daylight modeling, wind flow, computational fluid dynamics and operational carbon emissions. Elective courses on building science, environmental and life cycle accounting, and embodied carbon add additional methods that students apply to their projects.

Innovative environmental building design typically exceeds the capacity of standard performance analysis tools, so each team identified one research question to explore through physical prototypes, whose behavior was monitored and analyzed. Ultimately the question each team had to answer was how can they be confident that a particular bioclimatic design technique will work? Students were led through an initial process of idea testing, using basic concepts and calculations to confirm its potential. That was followed by a program of detailed digital and physical testing, optimization, and sensitivity analysis, so that student teams can demonstrate the effect of their design.

At a minimum, each team demonstrated the energy consumption of their project and the balance of renewable resources used to power the building.

Required Readings

Relevant readings are developed together by the class at the beginning of the studio and then further refined by each team as they narrow the focus of their work.

2022 Student Work Examples

Passive Survivability Living Lab

Students: XiaoXiao Peng, Xiaowen Yu, Zhen Lei (XXL)
Instructors: William W. Braham, Billie Faircloth
Teaching Assistants: Surya Prabhakaran, Max Hakkarainen

Team XXL focused on Passive Survivability, exploring the performance of the building during climate disasters when normal systems of power, water and food are disrupted. They worked at the scale of the individual unit, the building and local infrastructure to design systems that were both resilient and drew on local, renewable resources.

A modular façade element with integral photovoltaic panels was developed for the building envelope, which needed to be substantially replaced. The façade reduces the heating and cooling loads with insulation and shading, and provides power to operate the remaining systems in the building.

The building is already connected to a campus steam and chilled water loop, so the team proposed to localize the loop, creating a micro-grid for electricity and hot and cold water. By connecting the dormitory to buildings with different energy profiles and schedules, waste energy can be shared, using a heat pump system to transfer heat into the neighborhood loop and from a geothermal exchange system.

Community Adaptation Living Lab (CALL)

Students: Yun Gao, Urvi Pawar, Yingbo Liu, Xinyu Yu
Instructors: William W. Braham, Billie Faircloth
Teaching Assistants: Surya Prabhakaran, Max Hakkarainen

The CALL team focused on the role of community in climate disasters. They argued that "Community members with diverse perspectives and resources empower decision making in a crisis. The community adaptation living lab (CALL) was designed to encourage communal conversation. An informed and engaged community will increase its adaptive capacity as these community members collaborate and share ownership in the crisis."

The project advanced several strategies to enhance the dormitory community and its connection to the local community. A kitchen and dining area was introduced to bring the building occupants together. A publicly accessible cool gallery was conceived to bring the local community into the building and provide a valuable service in hot conditions. The team discovered that even with climate change there are enough cold days in the winter to freeze significant amounts of water. The ice can be stored around the building until summer to provide radiant cooling of select public spaces in the building and in shaded outdoor seating areas, providing an innovative summer gathering space even in extreme heat conditions.

Circular Economy Living Lab (CELL)

Students: Thamer AlSalem, Mickey Chapa, Ali Hashem, Shuqin Li
Instructors: William W. Braham, Billie Faircloth
Teaching Assistants: Surya Prabhakaran, Max Hakkarainen

Team CELL focused on the total material and energy economy of the building: "As building materials' extraction and manufacturing contributes largely to the detriment of our climate, the circular economy adds value throughout the value chain while drastically reducing production of new materials by introducing a leasing model and specific strategies that correspond to the type and condition of a material stock."

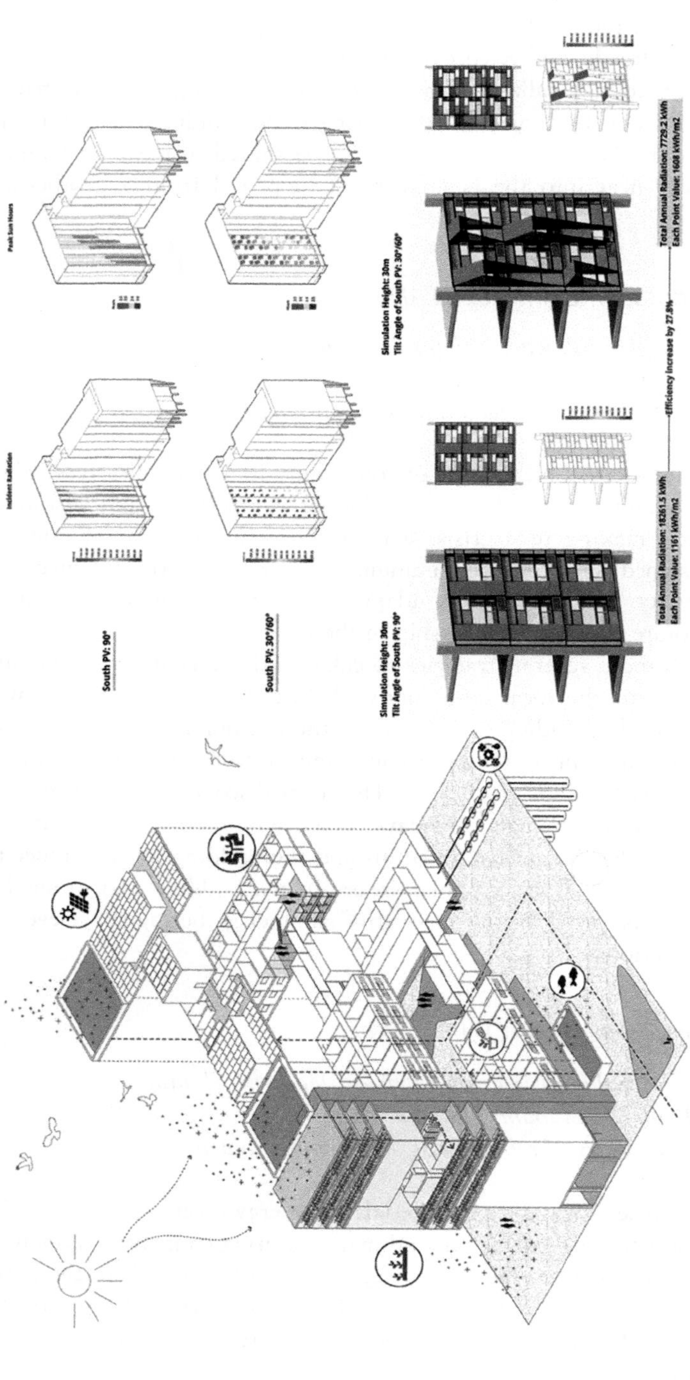

FIGURE 9.2 Passive Survivability Living Lab showing the total building system (left) and the modular façade (right).
Source: Zhen Lei, XiaoXiao Peng, Xiaowen Yu

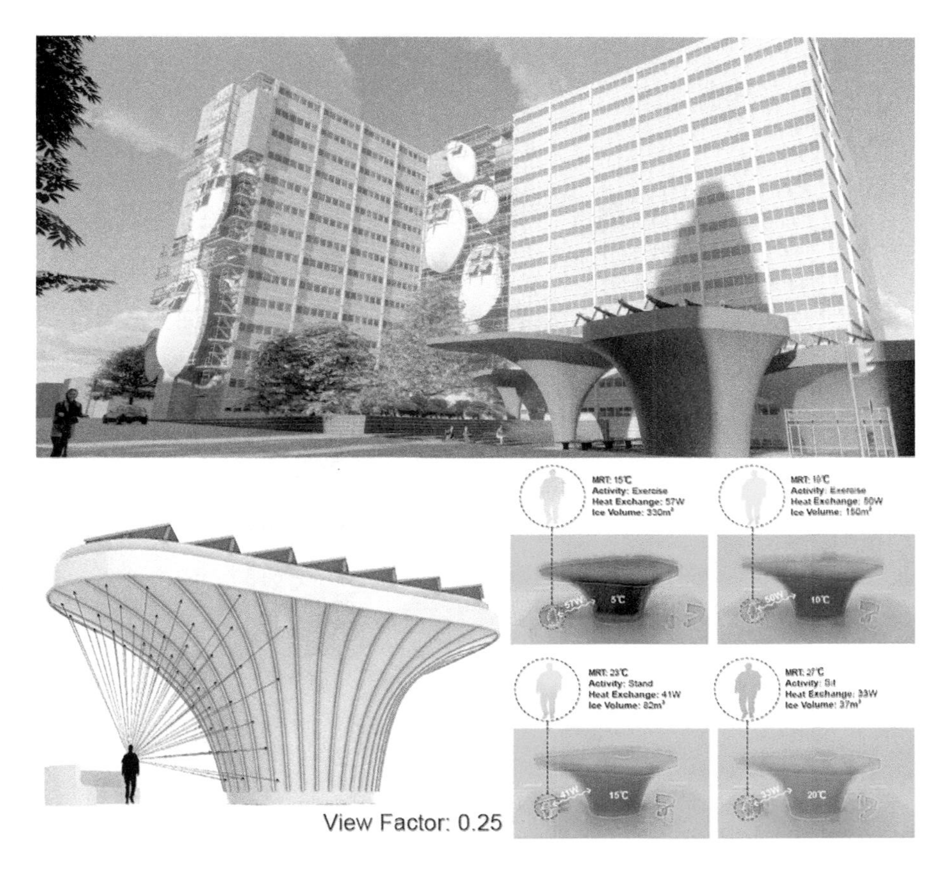

FIGURE 9.3 Community Adaptation Living Lab, showing external cooling pods and outdoor cooling zones (above), with climate chamber tests and view factor analysis (below).
Source: Yun Gao, Urvi Pawar, Yingbo Liu, Xinyu Yu

By providing various functional as well as educational spaces while maintaining its primary function as a dormitory, The Circular Economy Living Lab at Penn transformed Sansom West into a center for the evaluation and promotion of CE systems and products at all building scales and for all stakeholders.

Designing the multiple layers of architecture (Shearing Layers) for a climate emergency lends a powerful design framework to the project, including designing envelopes for disassembly, space plans for co-living, and stuff for sharing, while insuring all is done in a techno-environmental symbiosis."

The team prepared detailed material accounts for the existing building and sought to close the loops for all the elements that had to be upgraded, including a wood framed buffer space added to the south façade that employed multiple layers of single pane glass harvested from an adjacent renovation site.

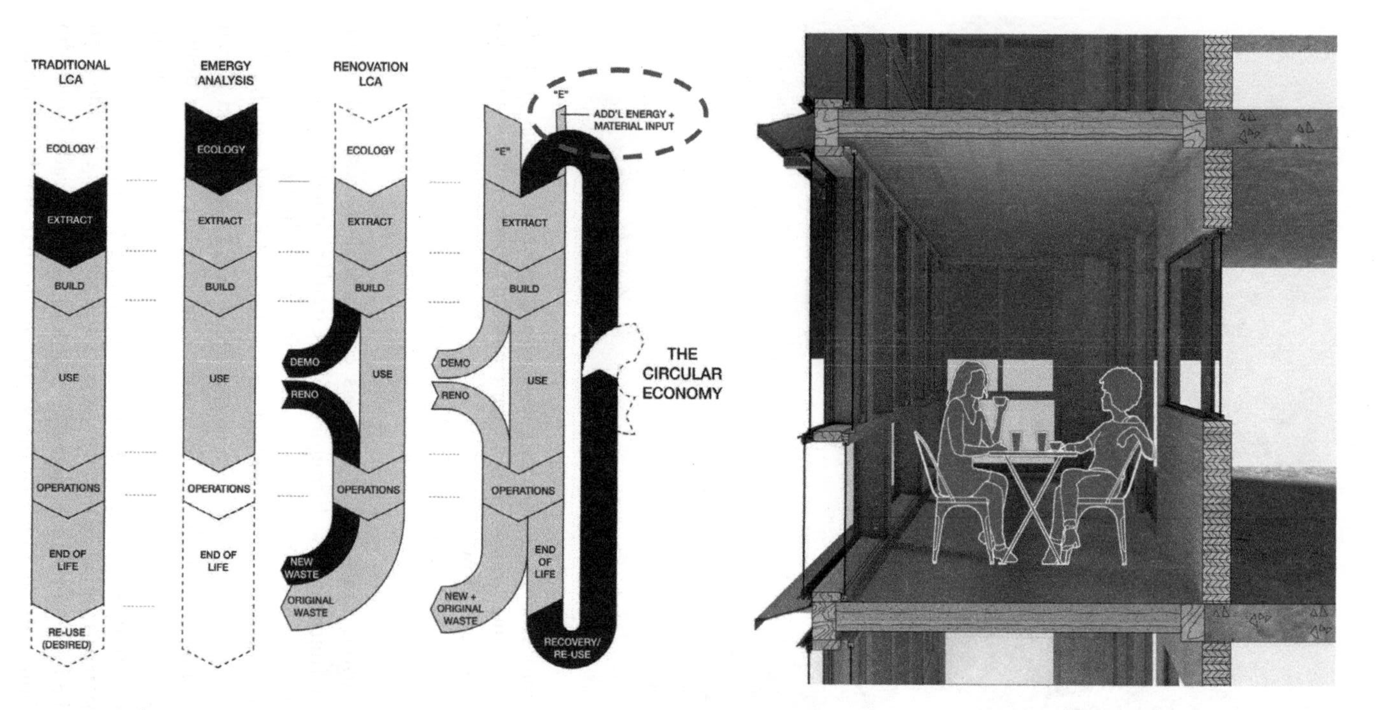

FIGURE 9.4 Circular Economy Living Lab, showing circular life-cycle analysis with emergy synthesis[3] (left) and wood framed buffer space addition with glass harvested from adjacent renovation site.

Source: Thamer AlSalem, Mickey Chapa, Ali Hashem, Shuqin Li

2023 Student Work Examples

Instructor: William W. Braham. TA: Max Hakkarainen.

SOLEVA // Solar Chimney with an Evaporative mass wall

Students: Hemant Diyalani, Ming Yan
Instructors: William W. Braham, Billie Faircloth
Teaching Assistants: Surya Prabhakaran, Max Hakkarainen

SOLEVA integrates an evaporative cooling mass wall with a solar chimney within the enclosure system of a desert shelter for two people to provide thermal comfort indoors. While the cooling loads are handled by evaporative cooling, the heating loads are addressed by redirecting warm air from the chimney back into the shelter. This dwelling spans a 36 square m carpet area, having a bedroom, living area, kitchen and a toilet.

The mass wall was extensively tested in the climate chamber and was complimented by two louvered buffer spaces used to regulate the conditions experienced by the wall. The roof structure used an integrated photovoltaics tile to provide the remaining power needed by the building.

FIGURE 9.5 SolEva // Evaporative mass wall (below) climate chamber prototype test (above). Source: Hemant Diyalani, Ming Yan

Desert Roof

Students: Dishanka Kannan, Ningxu "Janvier" Luo, Yujia Cui
Instructors: William W. Braham, Billie Faircloth
Teaching Assistants: Surya Prabhakaran, Max Hakkarainen

The project uses a Desert Roof, which connects an evaporative/radiant roof of sand with a rammed earth, radiant wall to keep the occupants comfortable in extreme heat. The roof can also be used to provide heat in the winter, using the same water loop to connect the roof to the radiant walls. By integrating these elements, the design optimizes thermal comfort and enhances energy efficiency. This innovative approach offers shelter in desert landscapes while reducing heat gain and fostering a sustainable solution for extreme climates. The desert roof not only provides a cooler environment in summer but also exemplifies a novel architectural direction.

The desert roof sits on a slope facing east and is approached from the west through a descending staircase. The roof protects the building from the harshest sun summer conditions but is positioned to allow the building to receive winter sunlight from the south. The water loop that circulates beneath the rooftop sand is circulated to an underground storage tank from which the cooled (or heated) water is circulated with the rammed earth walls to provide radiant heat exchange. The eastern edge of the roof supports an array of combined solar thermal and solar photovoltaic panels in vacuum tubes to provide the rest of the minimal energy loads of the building.

Smart Envelope for a Desert Glass House

Students: Zhirui Bian, Li Pan
Instructors: William W. Braham, Billie Faircloth
Teaching Assistants: Surya Prabhakaran, Max Hakkarainen

The team developed a desert shelter with a movable insulation system surrounding a glass house, capable of adapting to varying environmental conditions, especially hot days and cold nights. Vertical shutters insulated with vacuum insulated panels (VIP) respond to solar positions and temperature to regulate heat gain. The corner panels are translucent insulated with aerogel and along with a system of skylights provide daylight when the insulated louvers are closed. The interior glass envelope opens and closes according to temperature to facilitate natural ventilation and cooling when needed. Water stored in a roof pond further moderates temperatures and is also shaded by moveable louvers that allow it to capture heat on winter days and shed heat on summer nights. The lightweight prefabricated glass house is able to experience the best of its dessert location without the extremes of temperature.

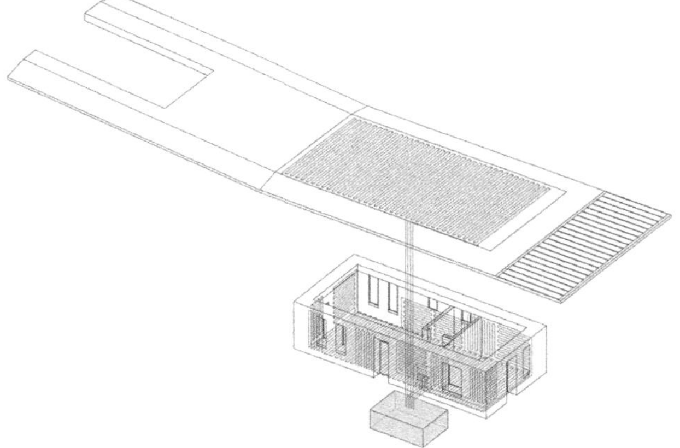

FIGURE 9.6 Desert Roof // View of roof (above) with diagram of evaporative surface and radiant rammed earth walls (below).
Source: Dishanka Kannan, Ningxu "Janvier" Luo, Yujia Cui

Outreach Opportunities

During the studio, students engaged selected stakeholders and subject matter experts. Students moderated a discussion with invited guests to explore adaptation and the role of human behavior, psychology, culture, equity and advocacy. The work is published on the program website, ebd.design.upenn.edu, and in *Pressing Matters*, the annual publication of the Department of Architecture.

Course Assessment

The student work is assessed by expert reviewers in the middle and end of the semester, including architects, engineers and building scientists. Students present and

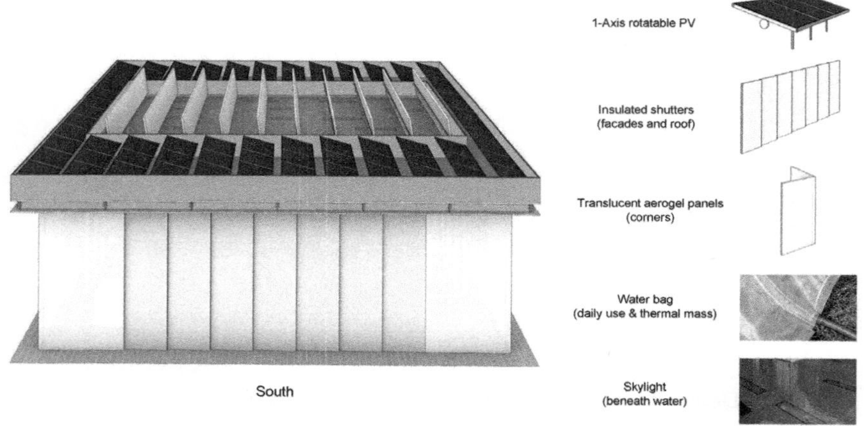

FIGURE 9.7 Smart Envelope // View of open louvers (above) and axonometric of components (below).
Source: Zhirui Bian, Li Pan

defend design proposals and the results of their performance analysis. The studio is also paralleled by ARCH 7550: Environmental Innovation and Prototyping, which helps students develop research plans to evaluate their design proposals. That course provides the opportunity to iteratively review and assess their analytical work so that claims about thermal, energy and carbon performance are well validated.

Looking Ahead

While the topic of the *EBD Research Studio* changes year to year, it continues to focus on environmental design innovation with prototyping and verification.

Based on five years of experience with the single-semester format, the studio will be extended over a full academic year beginning in 2024–2025. Student projects will be formulated and tested in the *Innovation and Prototyping* course in the fall and developed through the EBD Research Studio in the spring.

Notes

1 Victor Olgyay, *Design with Climate: Bioclimatic Approach to Architectural Regionalism* (Princeton University Press, 1963).
2 Carl Elefante, "The Greenest Building Is ... One That Is Already Built." *Forum Journal* 27, no. 1 (2012): 62–72.
3 Emergy synthesis is a comprehensive form of life-cycle analysis that accounts for all the work and resources dissipated in the production of a product, service, or activity, including the "free" work performed by global eco-systems. See Howard T. Odum, *Environmental Accounting: EMERGY and Environmental Decision Making* (New York: Wiley, 1996); and Howard T. Odum, *Environment, Power, and Society for the Twenty-First Century: The Hierarchy of Energy* (New York: Columbia University Press, 2007).

PART III

Performance Analytics

10

A RADICALLY TRANSFORMATIVE STUDENT-CENTERED APPROACH TO THE DESIGN OF NET-ZERO BUILDINGS

Robert Fryer and Rob Fleming

Course Overview

Course Name

SDN 622: Sustainable Design Studio

University / Location

Thomas Jefferson University / Philadelphia, PA, USA

Targeted Student Level

- Graduate
  - 2nd year, Master of Science in Sustainable Design, 2-year program

Course Learning Objectives

In this course, students will:

- Define the context for a design project from the lenses of scale, time and perspective.
- Define quantitative and qualitative goals for a design project, and identify strategies to achieve the goals.
- Create preliminary project designs using options as defined by different perspectives of inquiry.
- Create a schematic building design using passive design principles.

DOI: 10.4324/9781032692562-14

- Optimize a design project's energy efficiency by exploring different configurations of building envelope components.
- Integrate active technologies including photovoltaics, HVAC and water collection into a passively designed project.
- Analyze and communicate the performance of a design project against pre-established goals and benchmarks.

Course Learning Outcomes

Successful completion of this course will enable students:

- To think across time and space to design a project that addresses the global impacts of buildings and that also supports a sustainable future.
- To embrace ambiguity by developing multiple design options through different "lenses" (perspectives of inquiry).
- To transition from the outdated and hierarchical studio master/apprentice relationship to a self-confident and intentional design approach.
- To shift their understanding of design from a personal exploration in expression to an ethically foundational approach to design based in health, safety and welfare.

Course Methodology

Philosophy

The studio is informed and organized around the premise that sustainability is a new design paradigm for the built environment requiring explicit goals and metrics to which the designer must be held accountable to address the urgency of climate change. This leads to the primary innovation of a "cognitive flip" where achieving net-zero performance occurs prior to the traditional studio's priority of design expression (see Figure 10.1). Furthermore, the goals must be holistic to address the multi-perspective and complex character of true sustainability. Because sustainability is a new paradigm, the pedagogy must also adapt from the outdated hierarchical studio master/apprentice model to a transparent and authentic process that supports self-confident students empowered to propose design solutions for the health, safety and welfare of all via an ethically driven foundational approach. Architecture can shift from being a negative to a positive contributor to the environment by supporting a social-ecosystem-centric built environment.

Definitions

Carbon neutrality – Operational carbon emissions equal to carbon-free onsite renewable energy generation on an annual basis.

The Normal Design Process
(by nearly everyone)

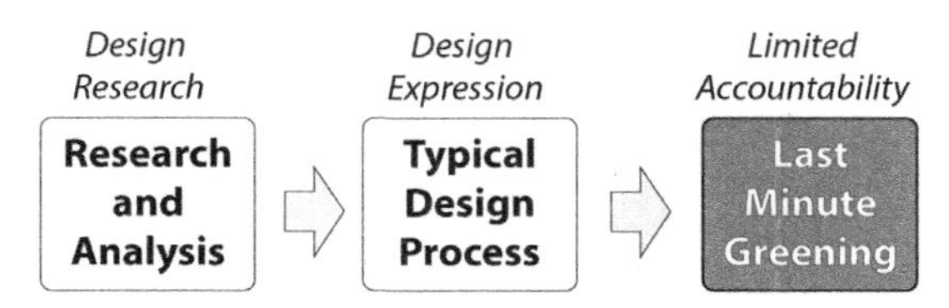

The Proposed Design Process
(By almost no-one)

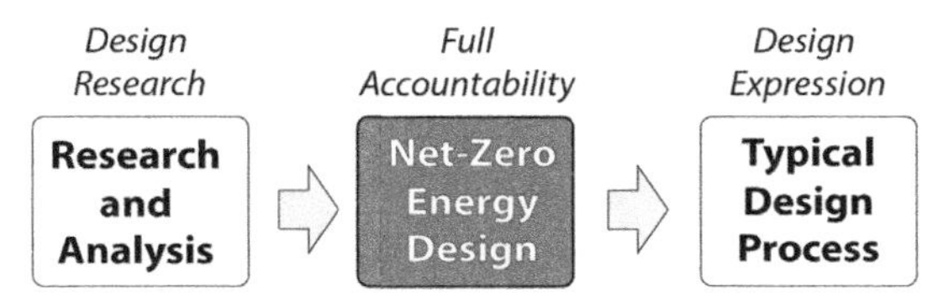

FIGURE 10.1 Net-Zero First primary studio innovation.
Source: Rob Fleming

Integral framework – The adaptation of Integral Sustainable Design used in the studio.

Integral sustainable design – A holistic framework developed by Mark DeKay based on the Integral Theory by Ken Wilbur, that unifies the subjective and objective aspects of a design project.

Net-zero vs net-zero ready – Net-zero is defined as a project that has energy consumption equal to onsite energy generation (ex. Solar PV panels). Net-zero Ready is the recognition that actual building use and occupant behavior may differ than the operational recommendations used in the simulation and the actual energy consumption may differ from the predictions.

Implementation

Studio Overview

First, in this section, a general overview of the studio process is provided, the key innovations of the studio, and then the steps (assignments) are discussed in more detail.

The studio uses a novel pedagogical approach that radically changes the normative design process to propose and test a net-zero building prior to other conventional aspects of studio, such as aesthetics and tectonics. This is the primary innovation of the studio, which is a transformational "cognitive flip" of

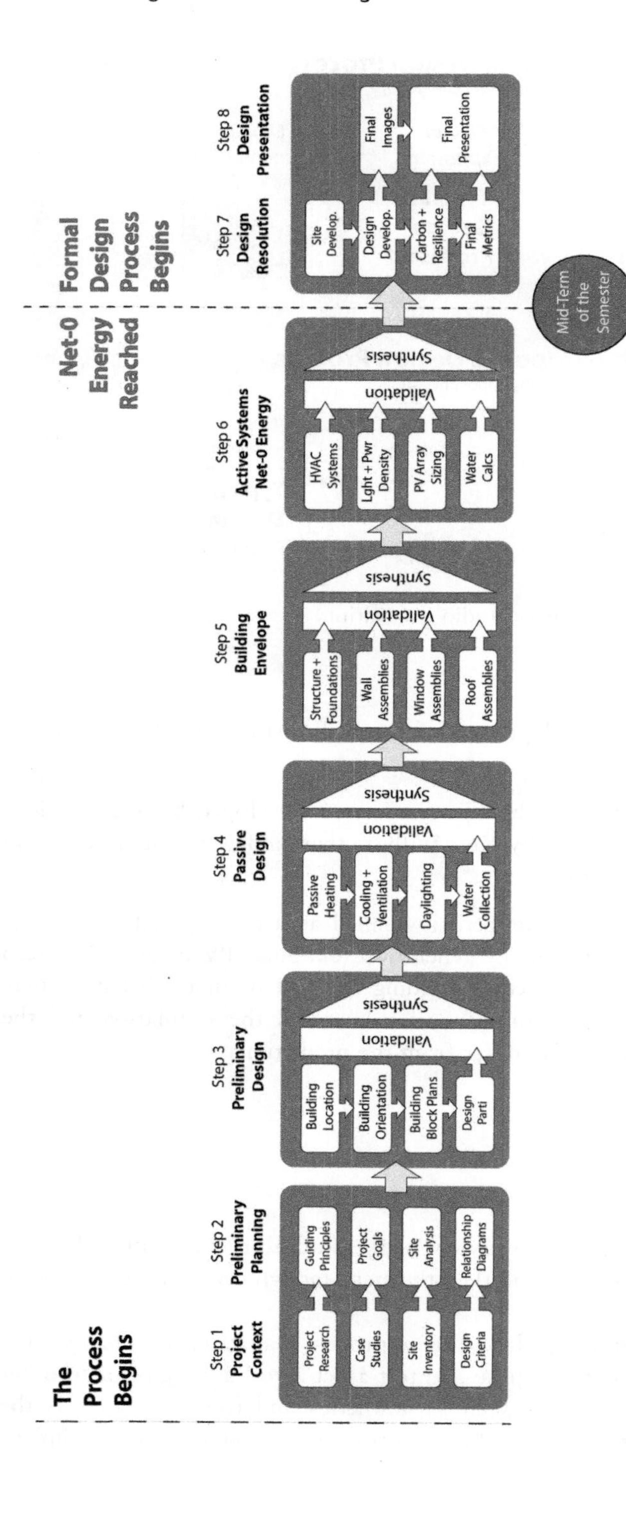

FIGURE 10.2 Studio organization.
Source: Rob Fleming

the design approach used in the traditional studio format. The studio is organized using a 10-step process of assignments that guides students through the novel design approach (see Figure 10.2). Students proceed through the course using a predesigned "template" to achieve the different learning objectives, thereby empowering them with confidence in the process and removing the mystery of studio schedules and faculty expectations.

The studio starts by identifying and understanding the project context, climate and program. This informs the predesign process by focusing research towards appropriate precedents to discover a range of strategies, goals and key metric outcomes. These are then evaluated for their potential application to their own projects, and students set their project goals and develop guiding principles. This is an essential and intentional exercise students revisit later in the course, post validation, to hold themselves accountable; an ethical priority sorely missing from the profession.

Students then explore options related to location, orientation, block plans and building forms to arrive at 16 different options that are validated using software for energy and daylighting potential, as well as against their guiding principles and design objectives. From these, one synthesis design is proposed and validated.

This "design loop" (the exploration of options, validation, synthesis and final validation) was tested and developed over time in multiple studios by multiple faculty in the MS in Sustainable Design Program, and forms the core design approach of the program. This iterative process includes the insightful application of the Integral Framework that generates four unique design proposals from different perspectives that together permit the discovery of design aspects that are otherwise ignored or forgotten in a normative approach, and instead lead to holistic solutions (i.e., the synthesis of four perspectives). This framework considers performance, but also other equally important components of a truly sustainable design (even for a performance-driven "net-zero studio"), such as experience of the spaces and impact on well-being, community inclusion and relationships, and goes beyond bioclimatic design to ecosystem integration and regeneration. It is used repeatedly in multiple steps in this studio.

For example, the iterative design loop is used again immediately in the next step with the application of passive design where the just-created synthesis model is again explored through multiple options, tested, synthesized and tested again. Students then propose and evaluate 16 options of structure and building envelope configurations, including exterior wall, window and roof permutations. After validating and synthesizing these into one proposal, the students have continually reduced their EUI without the intentional definition of an HVAC system, which happens next. The students undertake their final series of validation, by specifying and modeling their chosen HVAC system, electrical and lighting loads, and size an onsite renewable energy system to achieve (or surpass) net-zero. Water consumption calculations are also included, and savings noted. As their final step, students revisit their previously established project goals to hold themselves accountable, and compare their results to their performance, ecosystem, culture and experience goals to complete this larger "accountability design loop."

Ending the Tyranny of the Master/Apprentice Relationship

Other pedagogical changes besides the core design process and course organization contribute to the studio's innovative and radical approach. The first of these is the reconfiguration of the fundamental relationship between the faculty and students. Traditionally, this relationship was one of master to apprentice. This created a studio culture of fear, where the admiration of the seemingly capricious "master" was achieved by satisfying mysterious expectations, and the "apprentice" experienced intense judgment and competition for the "master's" attention, even from other students.

By contrast, in this studio professors serve as "mentors" to the students who are treated as "team members." The professor is less involved in the design resolution of the building itself and more focused on the broader executive skills of each student, thereby preparing them for practice. For example, students are required to perform a "test" presentation one week prior to the final review. Furthermore, the use of the template offers students a roadmap to completion, not unlike what they might find in an office. Typical studios are very "open-ended" in terms of how learning objectives are expressed by the student, relying more on a general drawing list. The open-ended approach has value in developing the student's ability to innovate and create a methodology and structure for a project. Instead, this studio is in response to the realization that students in a traditional architecture program need a mix of studio approaches, some prescriptive, like the Net-Zero-First Studio, and others with a more undefined process.

Replacing Inequitable and Ineffective Design Juries with Student- and Stakeholder-Centered Design Vetting

Another pedagogical revision is made to the design jury. In the past, the "jury" often resulted in an atmosphere of intense judgment, rather than an opportunity for students to learn by applying thoughtful expert feedback communicated in a constructive way. In the new approach, this has been replaced with design vetting, in which reviewers take turns talking to the students in a pre-defined structure and order that ensures a fair and equitable educational experience for every student regardless of gender, gender preference, religion, ethnicity or any other perceived difference. The format of the vetting process begins with any clarifying questions to ensure the design has been clearly communicated by the student. Next, the experts are asked to communicate what is working well in the student design, including optimisms about specific design moves, goals and the strategies used to achieve them. Then, the experts are invited to express cautions to the students, along with potential different considerations for the students to explore further. Again, the intent is to provide thoughtful feedback to enhance student learning and their design.

Shared Frameworks and Mental Models for Effective Sustainable Design

When traditional studios include sustainability, they lack a common language, framework and mental model. As a result, sustainable solutions aren't developed and evaluated continually nor comprehensively through an intentional process, and attempts at sustainability instead become a last-minute greening application without design proposals or strategies that are integral to the project. In such a process, carbon neutral design outcomes are difficult, if not impossible, to achieve. Instead, the Net-Zero First Studio establishes frameworks and creates new priorities at the beginning of the course which guide the design exploration consistently and comprehensively throughout the duration of the course. For example, the priority of the traditional studio is design excellence, with other components defined to support it, such as social equity and sustainability. By contrast, this studio shifts priorities to where design excellence is defined to be in support of sustainable design, along with environmental regeneration, social equity and performative design. Design excellence is now intentionally directed to achieve outcomes that eliminate carbon emissions.

As part of this reprioritization, other frameworks are deployed to achieve learning outcomes. The students become adept at using the Integral Theory Framework to define unique design options that respond to a multi-perspective reality of experience and emotion; culture and equity; systems and functionality; and behaviors and performance. This reality is understood across time and space, looking from history to the future, and across scales from the human to the global. Furthermore, the studio applies professional frameworks to organize precedent research to situate their self-defined goals among benchmarks. For example, students use the AIA Framework for Design Excellence[1] to impart a holistic approach to the project, and their precedent studies require students to collect information according to its categories.

Precedents

Students are asked to complete one case study as part of the preparation for the project. In this studio, the main goal of precedent study is to summarize the key learning outcomes from the project. Students enter performance metrics into a shared database of projects and use that information to situate their projects among precedents across the studio. Here, the Integral Theory Framework is aligned with the AIA Framework for Design Excellence to include energy, carbon, light, air, water, stormwater, ecology, resilience, diversity/equity/inclusion, community and experience (see Figure 10.3).

2.1 Case Studies **2.2.1 Benchmarks**

Sustainable Design Basics – Benchmark Matrix											
	Objective/Tangible/Measurable								Subjective/Intangible		
	Performance						Systems		Culture		Experience
Stated Guiding Principles	To showcase a commitment to sustainability by pursuing the Living Building Challenge (LBC)						To provide for passive survivability, resiliency, and adaptation		To create the gateway to Hampshire College, a commons and public gathering space for interaction between all members of the community		To bring people and nature back to the center of campus and to create a pedestrian-centered place
Metrics	**Energy**	**Carbon**	**Light**	**Air**	**Water**	**Storm water**	**Ecology**	**Resilience**	**Diversity, Equity and Inclusion**	**Community Walk Score**	**Views to the Outdoors**
	Predicted and **Net** Consumed Energy Use Intensity (Site EUI): kBtu/sf/yr NA	**CO_2 intensity** for building carbon footprint: predicted 54 kgCO2eq/m2	% of floor area or % of occupant workstations achieving adequate light levels or sDA	100 % of workspaces within 30 feet of an operable window	What percentage of water consumed onsite comes from rainwater capture?	% of rainwater managed on-site (from 24-hour, 2-year storm event)	% of landscaped areas covered by native of climate-appropriate plants supporting native or migratory animal	How many hours can the building function through passive survivability?	This metric is often controlled by the client and may involved the make up of the Design team, the stakeholders engaged or other metrics.	Subjective goals based on research and stakeholder engagement when possible	Percentage of floor area or percentage of occupant workstations with direct views of the outdoors
Actual Performance Results	SBEM Renewable Energy Ratio is 40%	embodied carbon associated with the structural frame of the building is over 50% lower when compared to a traditional hotel building SBEM CPC is 52 kgCO2eq/m2	% of floor area with daylight factor >2% **45%** %of floor area with daylight factor >5% **29%**	Air tightness test result achieved was 2.39 m3/m2/hr @ 50Pa	60% less consumption as compared to any other hotel	NA	80% of the site area	NA	Scale - 5	NA	100 %
Strategies & Tactics	100% of energy used on site is from renewable energy sources	Use of low carbon embodied materials.	All opaque elements were insulated to 0.15 W/m2k, and the punch windows and bespoke curtain wall glazing was 1.4 W/m2k	The ventilation system captures 81 per cent of rejected heat using a thermal wheel, which also heats the incoming fresh air for free.	Green roof to harvest rainwater	NA	Use of native plantation to attract birds and bees and maintain bio-diversity	no fossil fuels being consumed, the hotel also avoids producing pollutants in its local urban environment.	The architectural language of the project resembles the core values of Dublin's historical architectural context.	NA	All the guest rooms have an exterior view towards the street which brings in good daylight.
Synergies & Trade-offs	Net zero energy	Reduction in embodied carbon	Good daylight avoiding dependency on artificial lighting source	Appropriate ventilation system	Rainwater harvesting	NA	Promoting biodiversity	Use of renewable energy for all systems	Use of local architectural context and elements	NA	Merging indoor and outdoor boundary.

FIGURE 10.3 Studio precedent research matrix.
Source: Student work by Aditya Jahagirdar

Additionally, precedents serve the important role of developing executive functioning and the skill to summarize key information. This skill is often not developed in traditional studios. The precedent exercise replaces exhaustive case study research with short and powerful executive summaries, giving students the opportunity to complete, understand and summarize essential information to key strategies and tactics. These tasks are completed in a systematic application of the course template, adding another layer of analysis to move students from information collection to strategic application of their findings to set baselines for project metrics and to situate their projects within a continuum of existing projects.

Site Context: Community Relationships

Every design idea is required to express a community-oriented, equity-driven design solution. This requirement is included in each step of the "template" under the Culture and Equity quadrant. In this way, consideration of culture, community strengths, insight and support are integral to the design. This is a "codified" aspect of the studio design process that is a direct result of reimagining a new methodology to support integral sustainable design, as developed by Mark DeKay[2]. Having said that, the Net-Zero First Studio is primarily focused on the environmental aspects of a given design problem. This was intentional, as students have other studios that focus entirely on social equity, community engagement and responding to context.

Analytics

This, and other studios in the MS in Sustainable Design Program, continue to evolve their use and choice of analytical tools over time. The studio has a co-requisite course that applies various software for performance validation. Currently, the course uses a single plane SketchUp model to rapidly simulate projects using Sefaira for half of the semester before modeling their projects in Rhino to validate using Ladybug and Grasshopper.

There are tradeoffs with each software, however. Ladybug requires extensive coding, which affords students a richer educational experience regarding in-depth knowledge of energy principles and simulation technology. It can provide a portal to what is hidden behind the interface of "pre-packaged" software such as Sefaira. However, it takes longer to complete and often leads to student frustration as coding errors become obstacles to successful runs.

On the other hand, Sefaira is more facile at exploring rapid iterations, especially at early design phases, however it has limited options, such as building uses and shading.

The tradeoff between educational value, time and student frustration is continually evaluated and assessed. In teaching master's students in a university that prepares students for the profession, developing proficiency in software that is actually and most-frequently used in the profession must also be a driving force in the choice of analytical tools.

Assignments

This section looks at key steps (or assignments) in the studio.

Step 2 is the pre-planning phase. Here, students complete a precedent study to discover how specific metrics for experience, performance, ecosystem integration and culture are achieved. This information informs the benchmarks they set for their own project, and the students establish their project goals and the strategies to achieve them (see Figure 10.4). Later in the studio, they hold themselves accountable to these goals by comparing them against their final results. It's important to note that despite this being a net-zero studio, their goals are multidimensional and are expected to go beyond performance to achieve specific outcomes for the experience, community and ecosystem aspects of their project. Finally in this step, students create guiding principles that act as touchstones for design decisions yet to come in the process. The learning objective related to this step is: Define quantitative and qualitative goals for a design project and identify strategies to achieve the goals.

Step 3 is the preliminary design, where students explore multiple options for the location, orientation, block plans and building forms for their project. Students generate 16 options, and they are all tested for their energy use and daylighting potential (see Figure 10.5). Using these results, students propose one integrated design, validate it once more, and use that to develop further in the next step. The learning objective of this step is: Create preliminary project designs using options as defined by different perspectives of inquiry.

In Step 6, students identify the structure and building envelope components. Again, a multidimensional approach is taken, and students explore materiality from the selections that are appropriate for their context, to selections that maximize the performance, to ones that most benefit the environment, to ones that enhance the experience of their project (see Figure 10.6). Each of the explorations are again validated and tested for their impact on the EUI and sDA of their projects. Ultimately, one configuration is selected based on the outcomes of the validation, their guiding principles and design objectives. The learning objective for this step is: Optimize a design project's energy efficiency by exploring different configurations of building envelope components.

Step 9 is where students hold themselves accountable by revisiting the goals set in Step 2 and comparing them with the final results in Step 7 (see Figure 10.7). This is an intentional reflection to uncover why goals were or were not met and what might be done to achieve them. The learning objective for this step is: Analyze and communicate the performance of a design project against pre-established goals and benchmarks.

2.3.1 Goals and Strategies

Sustainable Design Basics											
	Objective/Tangible/Measurable								Subjective/Intangible		
	Performance						Systems		Culture		Experience
Proposed Guiding Principles	Buildings can produce their own energy on-site and reduce carbon emissions						Planning for catastrophes can help increase the overall well being of the occupants		Including the local community in the use and design of the site can help boost the local economy, improve health, and improve safety in the neighborhood.		An active neighborhood and site creates a safe and healthy neighborhood
Metrics	Energy	Carbon	Light	Air	Water	Storm water	Ecology	Resilience	Diversity, Equity and Inclusion	Community Walk Score	Views to the Outdoors or other quantitative
	Predicted and Net Consumed Energy Use Intensity (Site EUI): kBtu/sf/yr	CO_2 intensity for building carbon footprint: (lbs. CO_2/sf)	% of floor area or % of occupant workstations achieving adequate light levels or sDA	% of workspaces within 30 feet of an operable window	What percentage of water consumed onsite comes from rainwater capture?	% of rainwater managed on-site (from 24-hour, 2-year storm event)	% of landscaped areas covered by native of climate-appropriate plants supporting native or migratory animal	How many hours can the building function through passive survivability?	This metric is often controlled by the client and may involved the make up of the Design team, the stakeholders engaged or other metrics.	Subjective goals based on research and stakeholder engagement when possible	Percentage of floor area or percentage of occupant workstations with direct views of the outdoors
Proposed Goals	23/0	x	100%	100%	x%	75%	75% of site area	12 hours	scale 5	100/80/80	100%
Strategies & Tactics	Use of passive techniques	Low impact building materials	Use of daylighting	natural ventilation	Rainwater harvesting. Low flow fixtures.	Infiltrating stormwater into the ground	Use of native plants and trees to support biodiversity	Use of renewable energy sources for all systems	ADA Accessible. Community recreational areas.	Providing trails and pathways for better connectivity	Biophilic environment adapting to the on-site biodiversity
Synergies & Trade-offs	Net zero energy	Reduction in embodied carbon	Good daylight avoiding dependency on artificial lighting source	Appropriate ventilation system	Rainwater harvesting		Promoting biodiversity	Use of renewable energy for all systems	Use of local architectural context and elements	Fair inclusion of neighborhood in the design and use of the site	Merging indoor and outdoor boundary.

FIGURE 10.4 Example of student goals and strategies.
Source: Student work by Aditya Jahagirdar

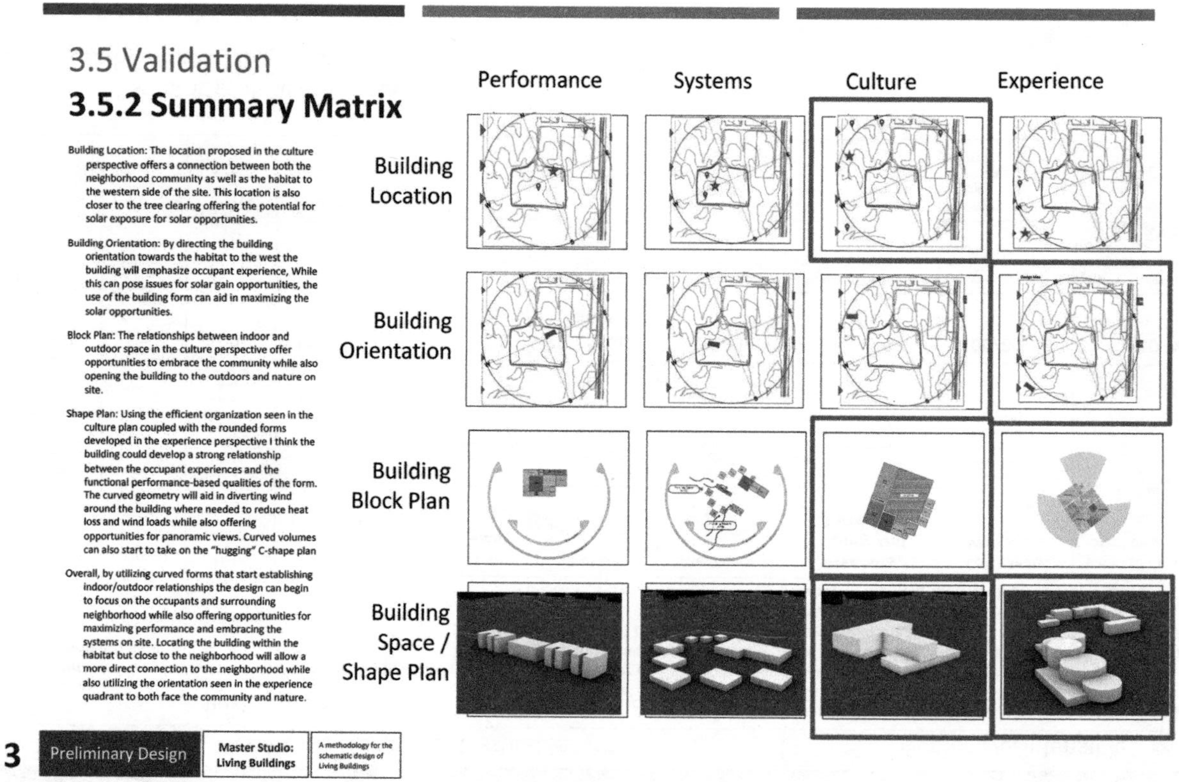

156 Teaching Carbon Neutral Design in North America

3.5 Validation
3.5.2 Summary Matrix

	Performance	Systems	Culture	Experience
Building Location				
Building Orientation				
Building Block Plan				
Building Space / Shape Plan				

Building Location: The location proposed in the culture perspective offers a connection between both the neighborhood community as well as the habitat to the western side of the site. This location is also closer to the tree clearing offering the potential for solar exposure for solar opportunities.

Building Orientation: By directing the building orientation towards the habitat to the west the building will emphasize occupant experience. While this can pose issues for solar gain opportunities, the use of the building form can aid in maximizing the solar opportunities.

Block Plan: The relationships between indoor and outdoor space in the culture perspective offer opportunities to embrace the community while also opening the building to the outdoors and nature on site.

Shape Plan: Using the efficient organization seen in the culture plan coupled with the rounded forms developed in the experience perspective I think the building could develop a strong relationship between the occupant experiences and the functional performance-based qualities of the form. The curved geometry will aid in diverting wind around the building where needed to reduce heat loss and wind loads while also offering opportunities for panoramic views. Curved volumes can also start to take on the "hugging" C-shape plan

Overall, by utilizing curved forms that start establishing indoor/outdoor relationships the design can begin to focus on the occupants and surrounding neighborhood while also offering opportunities for maximizing performance and embracing the systems on site. Locating the building within the habitat but close to the neighborhood will allow a more direct connection to the neighborhood while also utilizing the orientation seen in the experience quadrant to both face the community and nature.

| 3 | Preliminary Design | Master Studio: Living Buildings | A methodology for the schematic design of Living Buildings |

FIGURE 10.5 Design Matrix showing 16 options derived from multi-perspective exploration.
Source: Student work by Stephanie Jensen-Schmidt

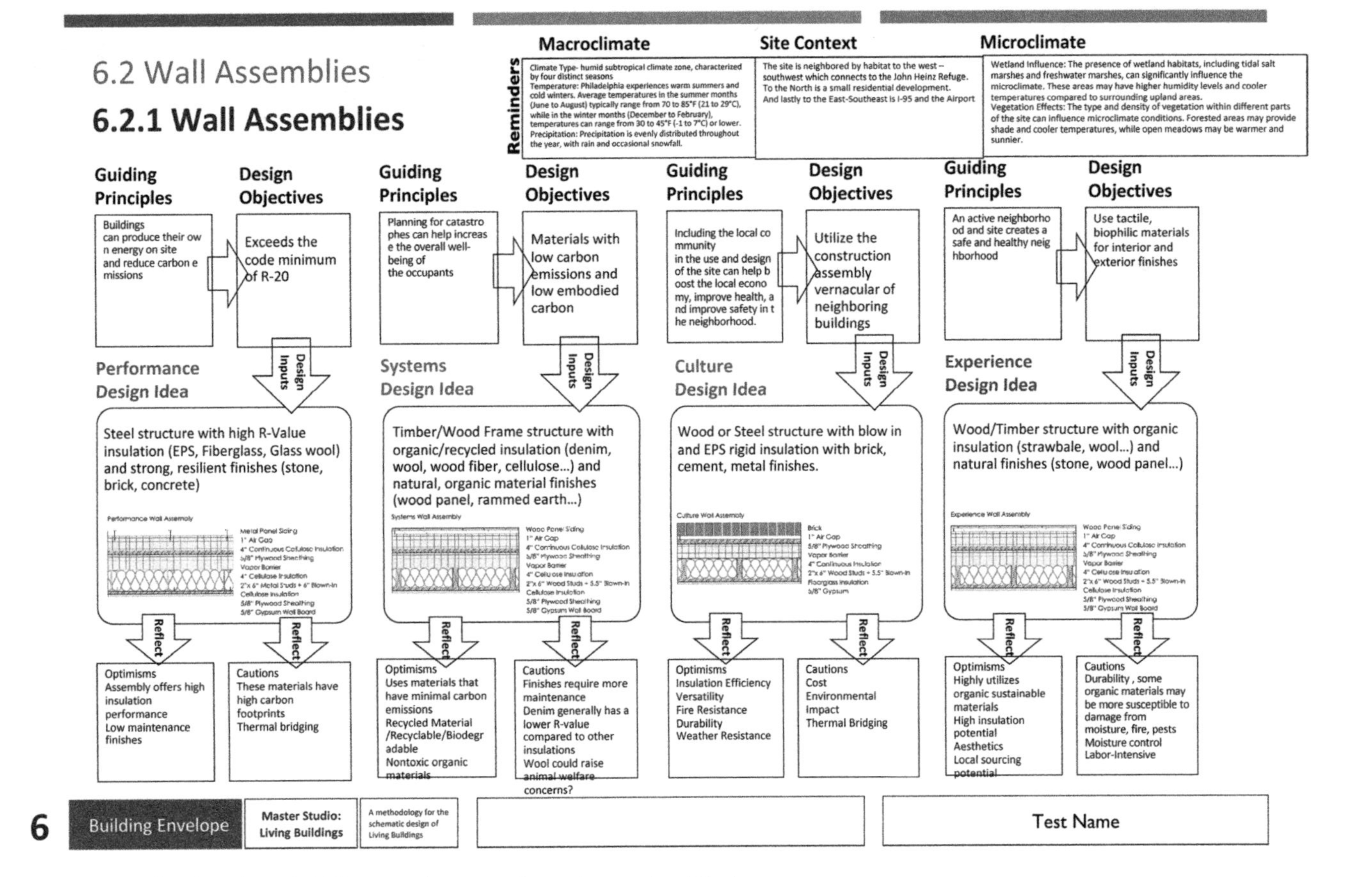

FIGURE 10.6 Example of student building envelope configuration and validation.
Source: Student work by Stephanie Jensen-Schmidt

9.7 Experience Goals

9.7.2 Summary

Category	Goal	Actual	Goal Met?
Energy	23/0 kBTU/sf2/yr	24.3/- 5.7 kBTU/sf2/yr	YES | Net Positive Energy
Daylighting	75% of Floor Area 75% sDA	99.5% of Floor Area 86.5% sDA	YES
Air Quality	100%	100%	YES
Water	25%	33.41%	YES | Net Zero Water
Habitat	100%	100%	YES
Community	Walk score of 80 Transit score of 75 Bike score of 80	Walk score of 70 Transit score of 70 Bike score of 80	NO NO YES
Experience	98% of Area with Views to Outdoors	99.5% of Area with Views to Outdoors	YES

9 Final Validation | Master Studio: Living Buildings

FIGURE 10.7 Example of students holding themselves accountable to their established project goals.

Source: Student work by Stephanie Jensen-Schmidt

Required Reading

Jaffe, Sharon B., Rob Fleming, Mark Karlen and Saglinda Roberts. *Sustainable Design Basics*. Hoboken, New Jersey: Wiley, 2020.

Outreach Opportunities

During the pandemic, this studio was taught online for free to anyone in the world that was interested. In the end about 12 students completed the entire course. It was a great test case to see if the specific net-zero-first approach was transferable to students who did not have a foundation in sustainable design. Anecdotally, the studio succeeded in engaging these students. The comprehensive template forced them to continue to integrate sustainable design techniques, strategies and metrics.

Course Assessment

The studio is organized around assignments with rubrics that were intentionally framed on each page of the design template. Professors grade the work based first on completion and second on the number of options generated. Aesthetic preferences for the design itself are removed from the grading rubric.

The template is also the format for their midterm and final presentations. This means that students did not spend any time thinking about how to present the project. The formatting of a presentation was not a learning objective for this course.

Assessment and grading occur weekly, leading to an overall grade for the course. Providing students with quick, objective feedback enhances learning and provides the opportunity to include feedback from one step in the next one. Furthermore, students always know where they stand in the studio. Their goal is to complete the project as per the learning objectives, not to make the professor "happy" with their design.

Looking Ahead

There are three areas of anticipated innovation. The primary focus of both academia and the profession must be the elimination of carbon emissions. As emissions associated with building operations begin to fall, the emissions associated with materials and their processes rise to the forefront and are the next challenge. Academia must develop the skill of embodied energy calculations in studios for students to lead the profession upon graduation. Embodied carbon calculation is already taught in another course in the program, but must be added to the studio to complement the current operational focus.

The second anticipated innovation is to incorporate biophilic metrics into the course. Biophilia, and mental and physical well-being more generally, are essential metrics when considering the professional ethical obligations to health, safety and welfare. More importantly, this focus will avoid the student's perception that sustainable design is only about EUI reductions and is, indeed, a holistic pursuit. Students currently receive lectures and readings on the topic of well-being, and even identify opportunities to include biophilic design interventions in their projects, but they are not yet held accountable for these decisions.

Finally, artificial intelligence is having an impact on the profession, and will have an impact on the studio, as well. Most significantly, it will lead to rapid analytics and assist in the validation of performative design options. At this time, the transformation that AI will cause is not clear, but it is expected to impact the software used in the course, and such changes will need to be integrated into the course.

Notes

1 AIA, 2023, *Framework for Design Excellence,* June, https://content.aia.org/sites/default/files/2023-06/Framework_for_Design_Excellence_June_2023.pdf, accessed December 1, 2023.
2 Mark DeKay, 2011, *Integral Sustainable Design: Transformative Perspectives*, Routledge.

11

INTRODUCTION TO CONCEPTUAL DESIGN PERFORMANCE ANALYSIS FOR CARBON NEUTRALITY

Lee A. Fithian

Course Overview

Course Name

Advanced Sustainable Resilient Systems

University / Location

University of Oklahoma / Norman, Oklahoma, USA

Targeted Student Level

- Undergraduate
  - 3rd year, 1st semester, Bachelor of Architecture, 5-year program
- Graduate
  - 1st year, 2nd semester, Master of Architecture, 2-year program

Course Learning Objectives

In this course, students will:

- Analyze climate data passive design strategies for a specific site using Climate Consultant and Architecture 2030 Palette to reduce energy usage.
- Estimate Daylighting Factors and what impacts them using software such as Andrew Marsh's web-based tool Dynamic Daylighting for daylighting performance analyses.

DOI: 10.4324/9781032692562-15

- Demonstrate Natural Ventilation Analysis using computational fluid dynamics (CFD) software such as the iPhone app Wind Tunnel for evaluating through-flow ventilation.
- Estimate simple Net-Zero Energy Building calculation targets to establish EUI and Decarbonization targets.

Course Learning Outcomes

Successful completion of this course will enable students to:

- Research and evaluate emerging science and methodologies for optimizing building and human performance.
- Use analysis, tools and methods to design, develop and assess the building systems and performance of decarbonized projects.
- Explain how design decisions impact society, systemic issues and the environment for all communities today and in the future.

Course Methodology

Philosophy

Architecture studios should make design decisions before a line is drawn. Site assessment has as much an impact on energy usage and operational decarbonization as selecting efficient HVAC equipment. In this studio, design starts with site analysis at the very beginning of the concept design using a variety of tools to interpret climate data, particularly Climate Consultant. Understanding the fundamental climate conditions of the site includes not just orientation and latitude, but prevailing winds and their associated seasonal directions, including the specific times of day or night that may provide cooling opportunities. It is also important to respect the climate data and not make assumptions, such as heat being a determinant of cooling systems yet not considering seasonal humidity levels. The site analysis will reveal the potential for passive opportunities, which the talented designer will employ as many as can be accommodated.

Concept-level daylighting design incorporates room-level daylight factor analysis that is further tested by natural ventilation analysis. The historic trend to analyze daylighting by checking sun angles does not meet the need for work surface daylighting levels nor does it provide sufficient information to restrict glare, unnecessary heat gain and light levels sufficient to offset electric lighting during daylight hours. A simple yet direct measurement used in daylighting analysis is achieving a daylight factor of 2.0 over 75% of the work surface. Andrew Marsh's Dynamic Daylighting simulator makes it easy to quickly model a space and determine whether the space achieves the desired performance target.

Daylighting performance analysis is further tested by using natural ventilation analysis. It is necessary to identify those windows that should be operable to allow for cooling winds to flush heated spaces or pre-cool to assist in meeting HVAC setpoints. Professional architects' basic designs imagine most natural ventilation using "magic arrows", presuming the flow of outside air to completely ventilate through a space. Comprehensive ventilation testing required completed models and wind tunnels of expensive computational fluid dynamics (CFD) simulations. Modern CFD apps such as Weber State University's Fluid Dynamics Simulation or Algorizk's Wind Tunnel for IOS can demonstrate using simple conceptual diagrams whether there are sufficient openings and displacement for true mixing and flushing with outside air. A daylighting simulation must be followed by a ventilation analysis to determine the proper design of the building envelope as well as the aesthetics of the façade. This is fundamental to providing the passive means necessary to achieve net-zero energy usage and carbon neutrality.

All these design decisions are normally accomplished far in advance of the full building concept design. Waiting until a whole building model is created to apply passive designs, or test daylighting and natural ventilation opportunities deters the designer from making impactful design changes. In addition, setting Energy Use Intensity (EUI) targets requires an understanding of the building potential presuming that the previous passive elements are applied. Using the Architecture 2030 Zero Tool allows for concepts to identify EUI targets and greenhouse gas (GHG) emissions, while their Zero Code Calculator estimates onsite renewable energy targets for a net-zero energy solution for operational decarbonization.

When all these simple – and free to the public – tools are added to the designer's decision-making at the onset of concept design, the design has more opportunities to reach a holistic design that profoundly limits operational carbon.

Definitions

Carbon neutrality – Most definitions of carbon neutrality are based upon operational carbon reduction, which in this studio is accomplished through the development of passive design followed by the overlay of renewable energy systems.

Computational fluid dynamics (CFD) – The use of numerical analysis and data structures to analyze and solve problems that involve fluid flows.

Daylight factor analysis – The ratio of the light level inside a structure to the light level outside the structure, accomplished using room-level tools to validate meeting work surface illumination levels as well as mitigating glare.

Energy use intensity (EUI) –Refers to the amount of energy used per square foot annually.

Greenhouse gas (GHG) – The Intergovernmental Panel on Climate Change (IPCC) defines greenhouse gases as those gaseous constituents of the atmosphere, both natural and anthropogenic, which absorb and emit radiation at specific wavelengths within the spectrum of thermal infrared radiation emitted by the Earth's surface, by the atmosphere itself, and by clouds.

Iterative design – The act of testing a concept using successive analytical tools and re-iterating the design until all the tools show an optimized daylighting and natural ventilation solution.

Natural ventilation analysis – The use of simplified Computational Fluid Dynamic (CFD) through flow at the room level to validate mixing with the chosen fenestration design.

Implementation

This studio starts with site analysis and then proceeds to precedent analysis. Traditional precedent analysis involves the study of existing building typologies and the selection of features relative to the current design. This studio also focuses on passive design strategies that have been successful in the building site's climate. Precedent studies of climate adaptive strategies are presented along with traditional precedent studies. During studio reviews, professionals in practice are invited to review the complete story of the concept development to offer critiques as to the implementation of the strategies in the overall design.

Analytics

Analytics in concept and design development are fundamental test points to the validation of the design objective during ideation. Rarely are physical models or drawings sufficiently expedient within the concept design phase as ideas are studied, tested, rejected, or modified in quick succession. Part of the hesitation to use older physical analytical techniques as well as newer comprehensive software tools is the interruption of the idea generation and design flow. In the past, hand calculations, model building and solar access testing interrupted concept generation and the techniques of analytical assessment fell by the wayside. More recently, comprehensive analytical tools such as *cove.tool*™, *Sefaira*™ and *Insight*™ require robust models that must meet the needs of the tool not the ideation flow of the designer. The advent of simple software-based analytical tools allows room-level concept analyses that can be used to validate the performance of the design idea.

Assignments

Climate Data Assignment

The first step toward integrating sustainable design into a project is to have an understanding of the site and its climate. Climate Consultant can produce

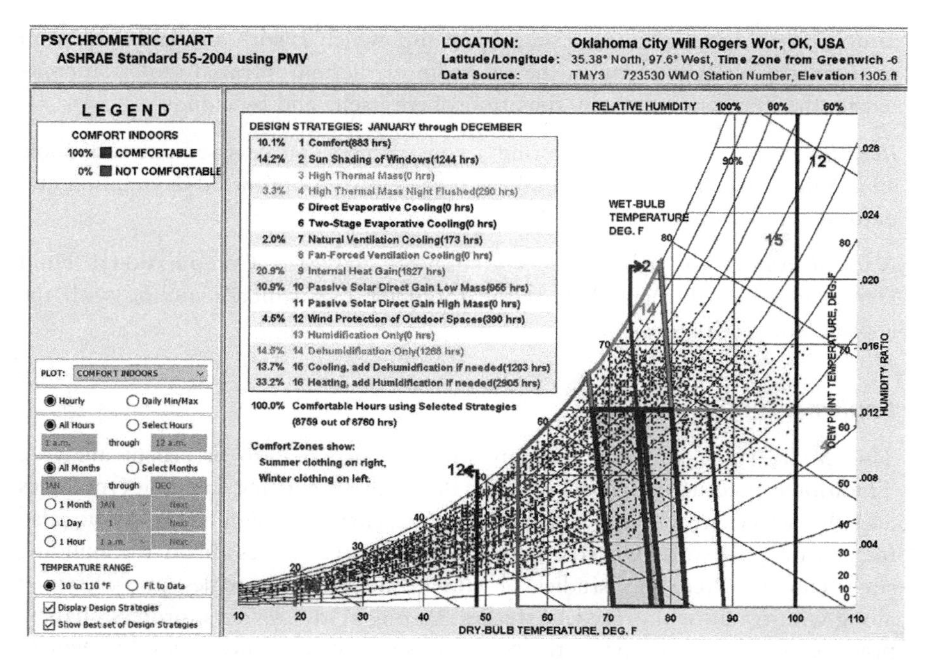

FIGURE 11.1 Psychrometric chart from climate consultant showing top passive design strategies for Norman, Oklahoma.

graphic results from climate data for specific locations and present this data quickly and efficiently. When you combine local climate data with your understanding of the Thermodynamic properties of dry and humid air, you can start to identify passive and active methods for heating and cooling spaces to provide a suitable level of comfort for the occupants.

After downloading and reviewing Climate Consultant in class, the students are asked to prepare a PowerPoint to present all of the following:

- Download EPW File for their studio location.
- Set Comfort Model to ASHRAE Standard 55 and Current Handbook of Fundamentals.
- Present the Temperature Range and answer the question: Is the Annual Mean Temperature higher or lower than the comfort zone? Does this surprise you? Why?
- Present the Psychrometric Chart with all Design Strategies.
- Identify the Top 5 Strategies (ie which 5 passive strategies have the highest percentage of hrs including #1 Comfort) and remove the remaining Strategies (those that have the lowest percentage of hrs) from the Psychrometric Chart and present.
- Select One Passive Cooling Precedent from the 2030 Palette and present.
- Select One Passive Heating Precedent from the 2030 Palette and present.

Measuring Daylight Assignment

From the Dynamic Daylighting online application:

> The aim of this app is to dynamically model the relationship between the spatial distribution of daylight in a room and its size, aperture configuration, shading devices, and external obstructions. To make this process interesting and fun to play with, the calculations are optimized to be as fast and responsive as possible without compromising on accuracy. This means that you can interactively change the room size, its surface properties, apertures, and shading devices, and the daylighting will update in close to real-time even on a tablet or phone.

Do a daylighting analysis of our studio understanding that the basic goal is to optimize daylight factor (DF) of 2.0% over 75% of the work plane along with the ventilation through the space is important. The assignment is to change the size of the room to one that is more like our classroom, which is a room that is 30 feet long (9150 mm) by 50 feet wide (15,250 mm) by 15 feet high (4750 mm). Notice that if most of the floor is purple, which is less than the desired DF of 2.0, that is because the windows are too small to bring in enough light. Also notice if the work surface is bright red, that indicates glare. It is useful to use shading devices as well as changing window heights and depths to optimize the daylighting factor of 2.0 over 75% of the spaces. *Assessment:* 100% – 75% of the floor must have a Daylighting Factor of 2.0 without more than 10% glare.

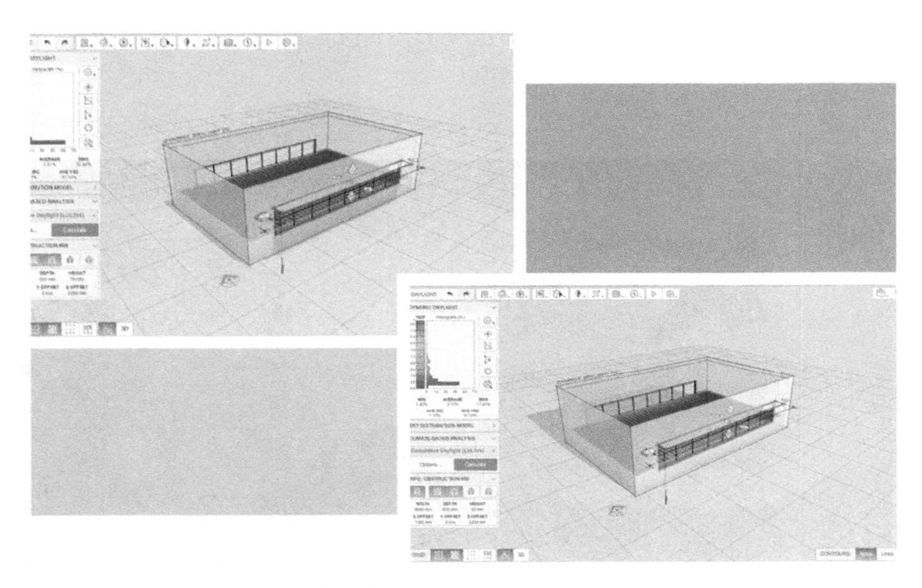

FIGURE 11.2 Daylighting model showing 75% of work surface achieving a daylight factor of 2.0 or more for studio space.

Natural Ventilation Assignment – Identification of Flow

Use the EPW for your studio space and set it to ASHRAE Standard 55.

Identify the Wind Roses for each month where *cooling* will be needed *or* where it is just *comfortable*. Take note where diurnal cycles may present opportunities for nighttime cooling. Prepare each slide so that they combine a site image (suggestion: use Google Earth), a Wind Rose and the temperature/humidity for that month.

Identify the Wind Roses for each month where *heating* will be needed (and ventilation rejected). Prepare each slide so that they combine a site image (suggestion: use Google Earth), a Wind Rose and the temperature/humidity for that month.

Use the studio plan layout that corresponds to the windows found in the Measuring Daylight Assignment.

Sketch the plan in the Algorizk Wind Tunnel App and show where windows need to be operable to get good natural ventilation and mixing during those times identified from the climate data.

Net-Zero Operational Energy Assignment

Now that the student has a site and a building orientation to optimize natural ventilation and daylighting, determine how the project can achieve a Net-Zero Energy Rating. Using your studio space enter the information into the Architecture 2030 Zero Code calculator and answer the following questions:

- How many square feet of roof do I have on my Project?
- Where can I put the PVs (collectors)?

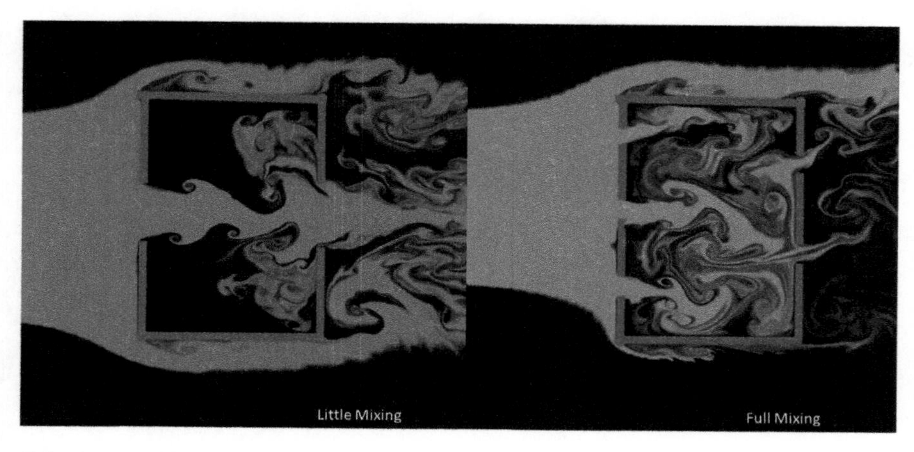

FIGURE 11.3 CFD image showing natural ventilation and mixing patterns in the plan of the studio space.

- If I don't have enough roof space, can I make a design change to get to the required number of square feet so I can power my building with PVs?
- If I don't have enough space and I can't make a design change, can I create some covered parking or use another site nearby to meet the total needed?
- Capture screenshots of your online input into the Zero Code Tool
- Show a generalized roof/site plan of your project showing the location of the PVs

Setting Target EUI and Predicting GHG Emissions Assignment

Using your studio space enter the information into the Architecture 2030 Zero Tool and answer the following questions:

- What is the Site Baseline and Target EUI (kBtu/ft²/yr)?
- What is the Source Baseline and Target EUI (kBtu/ft²/yr)?
- What is the Total GHG Emissions Baseline and Target (metric tons CO_2e/yr)?

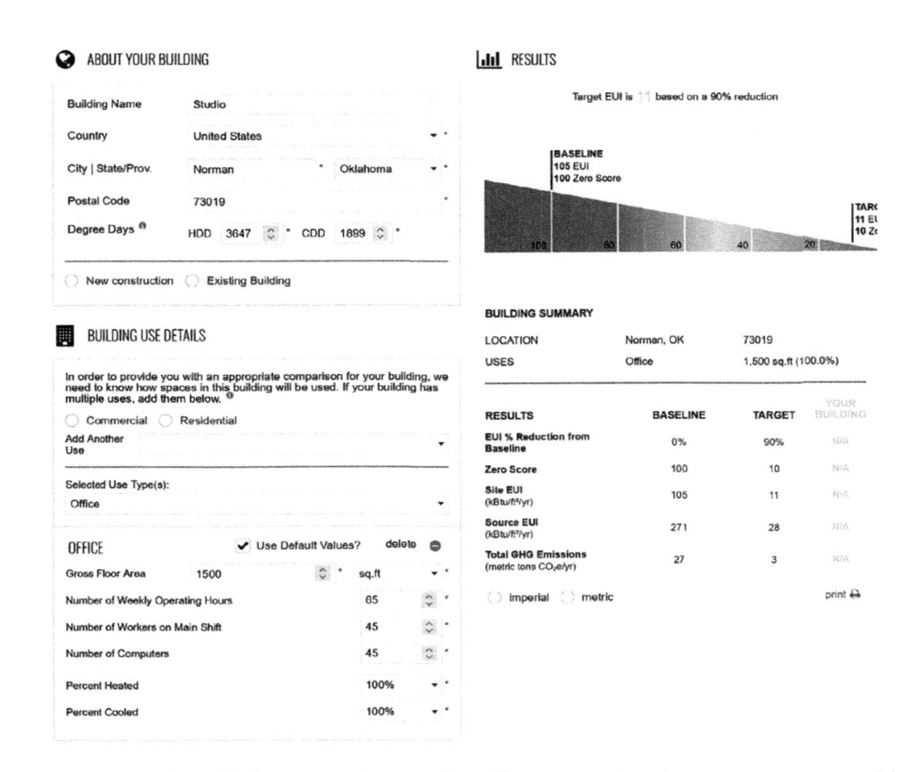

FIGURE 11.4 Zero Code output for studio office space showing net-zero renewable energy solution.

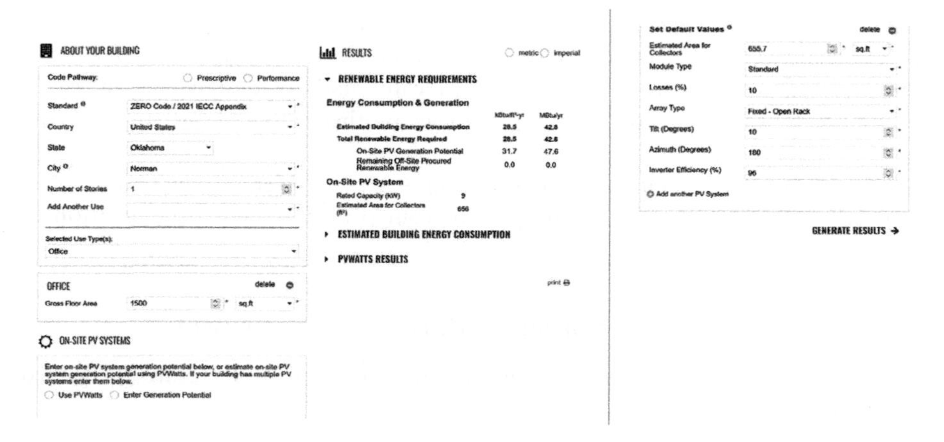

FIGURE 11.5 Zero Tool with output for studio office showing Target EUI and GHG emissions.

Required Readings

Arup, *It's Alive: A Vision for Tall Buildings in 2050*, www.arup.com/insights/its-alive-a-vision-for-tall-buildings. *It's Alive* is a conceptual vision for a "living" socio-technological tall building in the year 2050 that imagines tall buildings fulfilling multiple roles in response to the changing needs of our society and planet.

CADdetails, *How Many Buildings in the World are LEED Platinum Certified?*, https://caddetailsblog.com/post/how-many-buildings-in-the-world-have-become-leeds-platinum-certified. Out of the estimated 6,633 LEED Platinum buildings in existence today (and undoubtedly many more to come), we've highlighted 10 notable buildings that achieved a LEED certification.

Design! Life Depends on It [video], https://vimeo.com/101548831. Ed Mazria, Founder and CEO of Architecture 2030, delivered the keynote address at the AIA National Convention in Chicago in June 2014. The short video, titled Design! Life Depends on It, lays out the blueprint for a carbon-free and just-built environment by 2050, reviews the progress made in the building sector since issuing the 2030 Challenge in 2006, and outlines the critical role architects and designers must play in securing a livable future.

Juhani Pallasmaa, *The Eyes of the Skin*, John Wiley. First published in 1996, *The Eyes of the Skin* has become a classic of architectural theory and is required reading in courses in schools of architecture around the world.

Course Assessment

The hardest part for beginning designers is to develop their understanding of the iterative nature of design. This means not just the aesthetics, but the performance-based impacts. Once students see the need to reassess and change, and the methodology is incorporated into their design process, they become more accustomed to using the analytical tools to inform their design. This process must be carried forward through design critiques and subsequent studios before it becomes an integral part of their design process.

Looking Ahead

If carbon neutrality is truly a priority to the profession, then performance analysis tools should continue to be made available free of charge. The Daylighting Analysis in this chapter was originally developed using a Rhino plugin that was subsequently privatized. Although alternatives are now being used, multiple studios lost access to this methodology. Software has enabled conceptual frameworks to have performance analysis within the swift flow of ideation. Climate data and analysis must become predictive to account for ongoing global warming. Just 20 years ago, CFD analysis was performed on supercomputers now it is online and on iPhone apps. Daylighting analysis has reached a level where it is quick and simple to evaluate online and doesn't require a cumbersome setup and spatial definition. These simple daylighting and ventilation analyses must scale up to whole building modeling, and not hide behind a pay-per-use or subscription fee. The major building performance analytical tools should be freely available at the whole building scale and be as simple as the concept level room design. Without free and easy-to-use analytical tools, the software developers and corporations who build them are holding hostage the further decarbonization of the built environment.

12

REUSE STUDIO

Approaching a Carbon Neutral Future through Extending the Life of Existing Buildings

Omar Al-Hassawi and Kjell Anderson

Course Overview

Course Name

ReUse Studio: Approaching a Carbon Neutral Future through Extending the Life of Existing Buildings

University / Location

Washington State University / Pullman, Washington, USA

Targeted Student Level

- Graduate
  - 1st year, Master of Architecture, 1-year accelerated option program
  - 2nd year, Master of Architecture, 2-year program
  - 3rd year, Master of Architecture, 3-year program

Course Learning Objectives

In this course, students will:

- Analyze the context at multiple scales beginning from the immediate site up to the surrounding neighborhood and region through a site visit, site interviews and readings developed by the City of Post Falls which outline pathways for its future growth.

DOI: 10.4324/9781032692562-16

- Learn a comprehensive approach to carbon neutral design through discussions, readings and workshops that cover the AIA Framework for Design Excellence, the Path to Zero Carbon Series and building performance analysis software.
- Become familiar with terminology used in describing the different approaches to reusing existing structures through readings, and discussions that analyze national and international built examples with unique approaches to adaptive reuse.

Course Learning Outcomes

Successful completion of this course will enable students to:

- Create a comprehensive contextual analysis and use its outcomes to support a design narrative, influence programming and drive building massing. Students review the history of the site and the greater Post Falls area dating back to the time preceding when it was founded. They understand the veneer manufacturing process that previously occurred on the site and within the existing structures described in Figure 12.1.

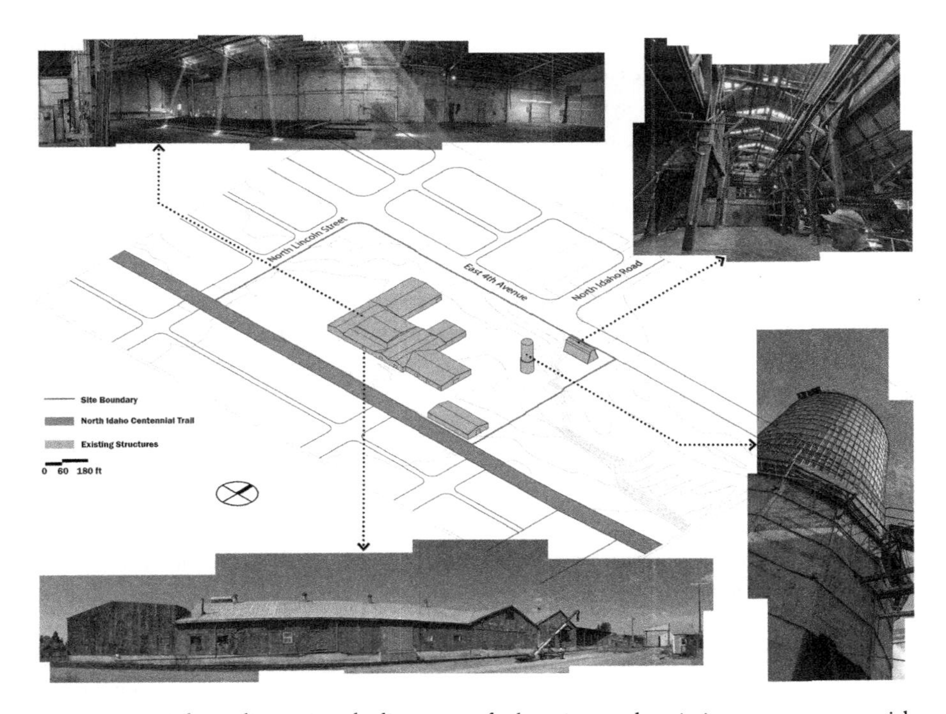

FIGURE 12.1 Three-dimensional diagram of the site and existing structures, with photos of the main structures remaining on site, taken in August 2020.
Source: Omar Al-Hassawi

Students analyze its spatial requirements which influenced the growth of the veneer mill over the years. Macro level analysis covers site demographics and findings from the Post Falls Comprehensive Plan, Community Center Feasibility Study and the Downtown Public Space Feasibility Study. Regulatory documents are reviewed to understand minimum requirements including the Post Falls Smart Code, the fire and life safety code, and the energy code. Finally, students understand the climate conditions in the region and the natural resources arriving on the site to better integrate them into the project.

- Design a master plan that effectively repurposes the existing structures while simultaneously introducing new construction to meet the population density requirements outlined in the Post Falls Smart Code. The proposed programs must draw from the analysis, address client needs and activate the site in a way that provides opportunities for the residents of Post Falls to invest in their city and change its common perception of being a bedroom community for the cities of Spokane and Coeur d'Alene. The proposal in Figure 12.2 responds to Post Falls' rapidly growing population by connecting the expansive site through residential, commercial and art co-op typologies distributed on the site into three overlapping sectors. The co-ops help invest in the local community by creating new job opportunities while providing high-quality living spaces on site.
- Integrate building systems into the design that adapt to the existing structures, respond to the site's climate conditions and maximize the use of the natural resources arriving on site. The structural systems of the new programs must respond to the existing structural layout when placed into the existing buildings. Passive environmental control systems compatible with the climate are incorporated in a way that affects the shape of the buildings. High-performing enclosure assemblies must exceed minimum code requirements and use materials that are locally sourced. Students use building performance analysis software to assess the long-term environmental and economic impacts of their design decisions, including embodied carbon, annual operational energy consumption, annual indoor daylight quality and on-site water management. Figure 12.3 is a project that maintains the structures of the veneer mill as an integral part of the site's future by building new spaces above the existing structures using a carbon sequestering mass timber structural system aligned with the existing structural grid. This creates the opportunity for a variety of pathways through the existing buildings with unique experiences on the ground floor, all of which converge at the Confluence, the epicenter of the project which sits at the lowest elevation on site and along the Centennial Trail. The inspiration for the project in Figure 12.4 is the semi-subterranean communal pit house developed by the People of the Plateaus which

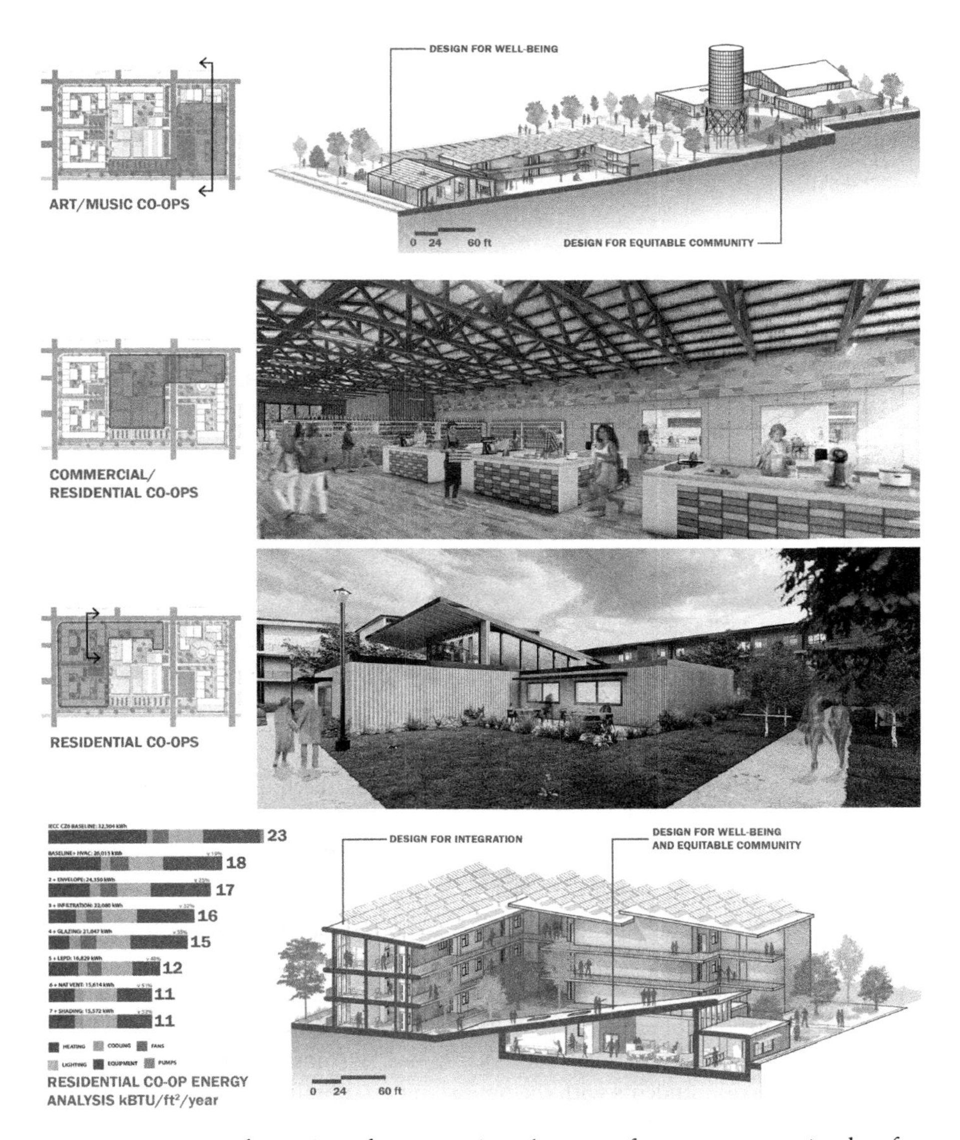

FIGURE 12.2 Example project demonstrating the use of co-ops as a stimulus for growth. Illustrations: north–south sections, renders, diagrams and annual energy reduction analysis for one residential co-op.

Source: Lauren Pate, Gracie Priddy and Taryn White

responds to the local heating-dominated climate and creates thermal insulation by redistributing the disturbed earth over the walls and the roof of the new construction. In this project, the soil excavated for the heat trapping sunken bath house is reused on site as thermal mass and green berms,

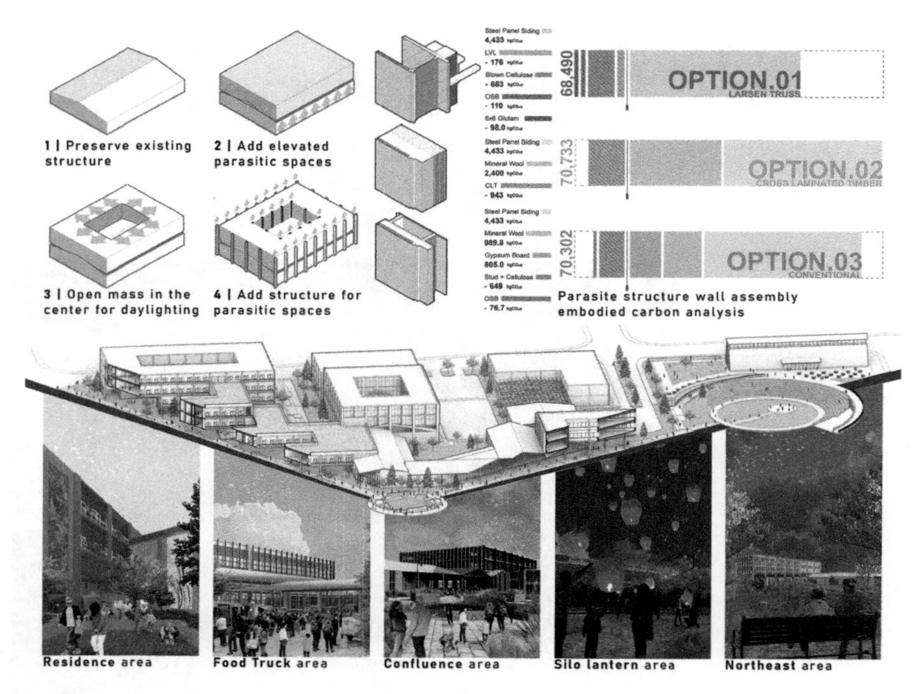

FIGURE 12.3 Example project demonstrating aligning new structural systems with the existing structural grid. Illustrations: east–west section, renders, diagrams and embodied carbon analysis for the different wall assemblies tested for the elevated structures.

Source: Zach Colligan, Abisai Mendoza Arroyo and Anthony Noble

significantly reducing the embodied carbon associated with soil transportation away from the site.

- Communicate design concepts using several techniques such as sketches, diagrams, orthographic drawings, perspective renderings, photography, physical models, design performance modeling and writing. Internal and external critics provide students with the opportunity to obtain feedback on presenting the narrative as well as refining the proposals visually and verbally.

Course Methodology

Philosophy

The greenest building … is the one that is already built.

– *Carl Elefante*[1]

Old ideas can sometimes use new buildings. New ideas must use old buildings.

– *Jane Jacobs*[2]

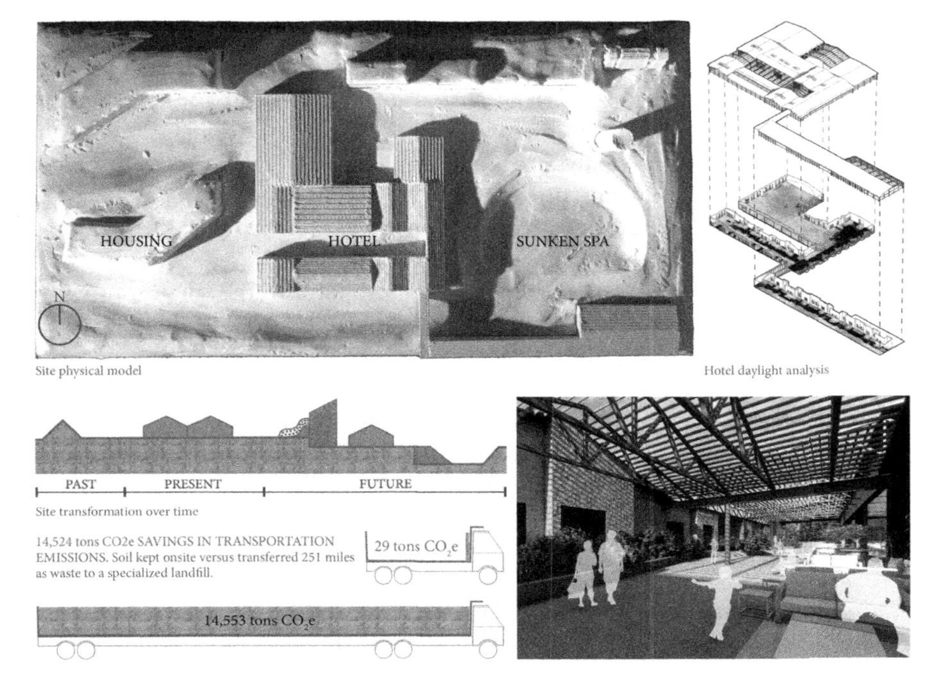

FIGURE 12.4 Example project demonstrating reshaping topography to implement passive heating strategies historically present in the context. Illustrations: physical model, render, daylight analysis and embodied carbon analysis for soil relocation.

Source: Jean Baker, Yuen Lei Lam and Sweta Waiba

The primary question the students are challenged with in this studio is: How can the life of existing structures originally designed for industrial occupancies be extended by introducing new programs which align with the future needs of their surrounding context?

Engaging students in a design process whereby structures on a site must be retained/repurposed is a necessary experience for future professionals in the field of design and construction since nearly 45% of design activity they will be involved in after graduation will be devoted to improving existing buildings[34]. The selected site and context allowed for collaborating with leading experts in the field of carbon neutrality who brought the underlying theories and assessment methods into the classroom through a series of discussions, workshops and Q&A sessions. Partnering with a construction firm, which gives students access to an actual project they are working on, is a key learning vehicle for this studio. Students gain the opportunity to generate proposals for a site currently under development by a regional firm, engage in discussions with an actual client, obtain feedback on their proposals from professional practitioners and network with potential future employers.

Refining design decision-making in this course is done iteratively. In addition to the testing and analysis of architectural criteria, such as contextual responsiveness, structural organization and spatial quality, students use building performance analysis software to evaluate their design's operational patterns and environmental impacts. Analysis of architectural criteria and building performance are neither linear nor prescriptive. Both require testing, modifying and retesting of several iterations to have a meaningful contribution to the creative feedback loop. Students test building performance hypotheses in tandem with the conceptual design origins of their building proposals rather than waiting for the creative process to end so that the technical analysis process can begin.

In the iterative design process, the key intents driving design proposals are maintained as more layers of detail are introduced to the project. The intents are consistently referred to as the project progresses into new scales of development beginning with the overall massing at the start of the semester and ending with the enclosure assembly details in the final phase of the semester.

Definitions

Carbon neutrality – A product or building for which any/all greenhouse gas emissions that supported creation and operation of it have been removed from the atmosphere. Carbon neutrality claims need to be comprehensive or state their limited scope.

Adaptive reuse / reuse – Maintaining rather than removing an existing structure(s) on a site and repurposing it with a function other than what it was originally intended for. Environmental impacts associated with the creation of the original structure are preserved, thus minimizing the built environment's impact on the natural environment.

Building performance analysis – The process of assessing how well a proposed design functions and/or impacts the natural environment early in the design process, primarily using computer software which operate within 3D visualization software ubiquitously used by design and construction professionals. Software introduced in workshops assesses the environmental impacts over the life of the building, the annual operational energy impacts and the annual indoor daylight qualities.

Implementation

General

Students in this studio work in teams of two to three and go through four phases to complete the semester project.

Phase One: Site Analysis + Existing Building Programming lasts for three weeks where students visit the project site as well as collect data about the site and context from online and print sources. They meet the former owners of the veneer mill as well as the current developers of the site. Student teams critically analyze the project site in terms of history, context, climate, resources and codes. Each team is assigned one area to focus on, thus minimizing repetition and overlap. A first iteration of the program in the existing buildings is proposed by the teams at the end of this stage justified by outcomes from the analysis. Students document their outcomes on presentation boards and present their findings to faculty and the overall class. Readings covered include the City of Post Falls Smart Code, Comprehensive Plan, Community Center Feasibility Study and the Downtown Public Space Feasibility Study (*see Required Readings Section below for each of these phases*).

Phase Two: Existing Building Development occurs between weeks four and six of the semester. At the beginning of this phase, student teams analyze adaptive reuse projects to draw inspiration and become familiar with the different categories used in describing the forms of introducing new programs into existing structures. Teams then develop a vision for the existing buildings and communicate that vision through a refined spatial program, modifications to the building massing and the intended overall design language. At the end of this phase, students present their progress to faculty and external reviewers using digital slides. Software workshops and meetings with industry experts advance the concepts and quantitative analysis of building performance and carbon neutrality. Readings include the Path to Zero Carbon Series and the book titled *Old Buildings New Forms*.

Weaving with the existing structures and the surrounding community spaces is used as an organizational strategy in the proposal shown in Figure 12.5. A series of interconnected sky bridges that link various programming spaces together weave in between the steel trusses and interact with public spaces located in the existing buildings and along the east edge of the site. Additionally, sky bridges reach out to new apartment buildings along the west edge of the site, thus stitching the public with the private. To create a stronger sense of community, the site's public programs complement the pre-existing gathering, creative and green spaces in the surrounding context. Inserting new programs into the existing structures is the organizational strategy for the project in Figure 12.6. The center of the site where most of the existing structures are located is dedicated to small-scale commercial spaces with a variety of sizes and material finishes taking inspiration from Idaho's small towns and aligning with Post Falls' goal to keep their small-town quality. Larger-scale residential buildings surround the commercial spaces and offer incubation spaces for business to expand into in the future.

Phase Three: Master Plan Development takes place between weeks seven and ten of the semester. Student teams address comments from phase two and continue to refine their vision by zooming out and iteratively developing a comprehensive master plan with attention given to the outdoor spaces surrounding existing and newly proposed structures. The same depth given to the

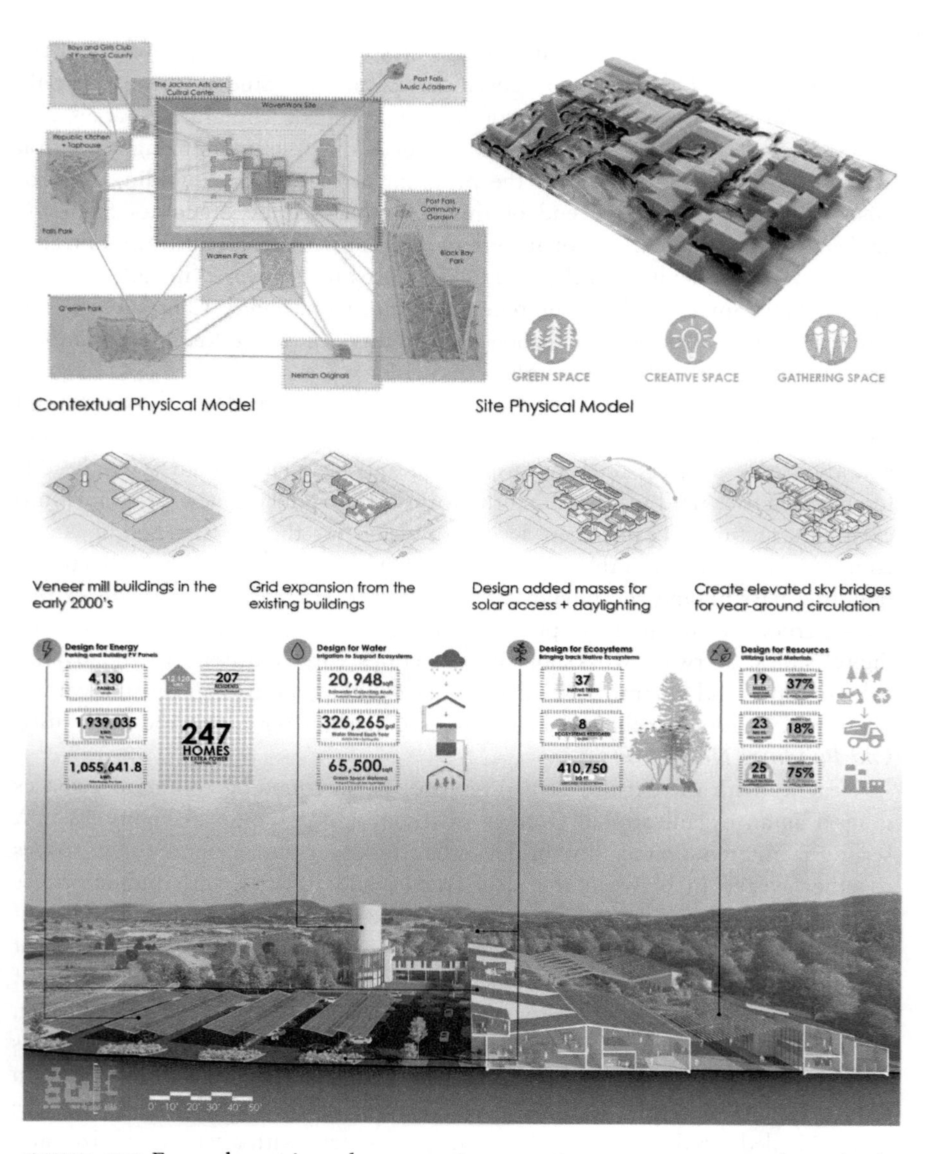

FIGURE 12.5 Example project demonstrating weaving new programs through the existing structures. Illustrations: site physical model, context physical model, diagrams and north–south section with calculations of energy, water, vegetation and wall assemblies embodied carbon.

Source: Rae Hendricks, Kaitlyn Maines and Dakota Witte

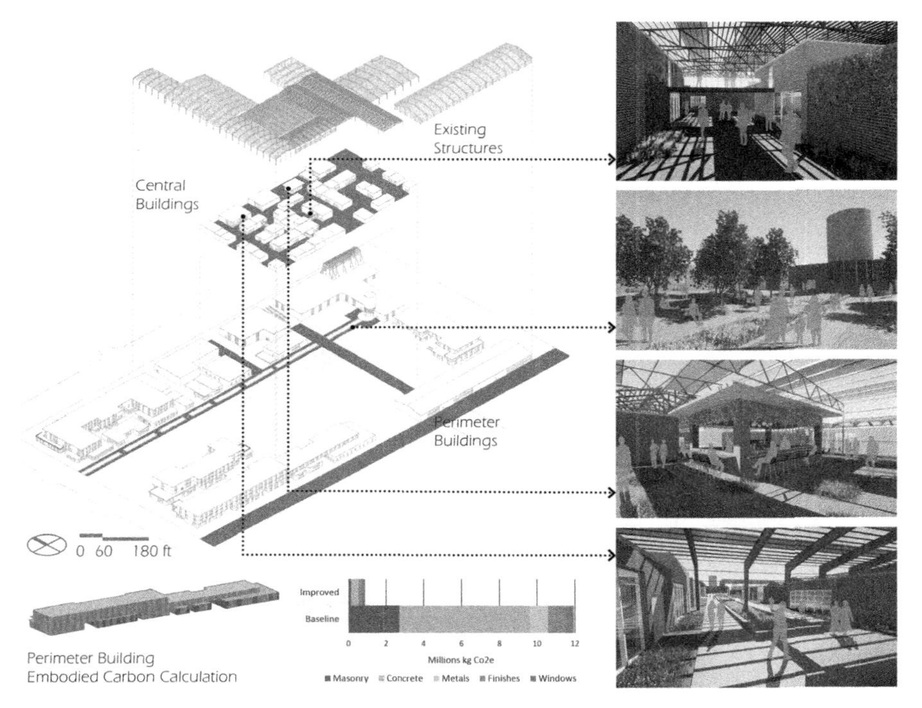

AQ3

FIGURE 12.6 Example project demonstrating inserting new small-scale programs into the existing large-scale structures. Illustrations: exploded axonometric, renders and embodied carbon calculation for one perimeter building.
Source: Mallory McKendrick, Anna Scott and John Sefuentes

indoor experience is given to the outdoor experience, particularly circulation within the site and beyond connecting to the surrounding context. Students begin describing the application of a subset of the AIA Framework for Design Excellence that most aligns with their overall scheme and produces their first results from building performance analysis software. All massing and internal layouts for the new buildings on site must be intentional even though architectural detailing will occur on a portion of the project due to the large scale and limited time. Software workshops and meetings with industry experts continue to occur in this phase to further the concepts of building performance analysis and embodied carbon. Students present their progress to faculty and external reviewers at the end of this phase through digital slides, two presentation boards and one physical model (1/128″ = 1′ – 0″). Readings include a continuation of the Path to Zero Carbon Series and selected chapters from the books titled "Re-USA" and "Walkable City Rules."

Figure 12.7 is an example master plan that inverts the indoor spaces of the veneer mill into outdoor spaces. It frames the perimeter of the existing

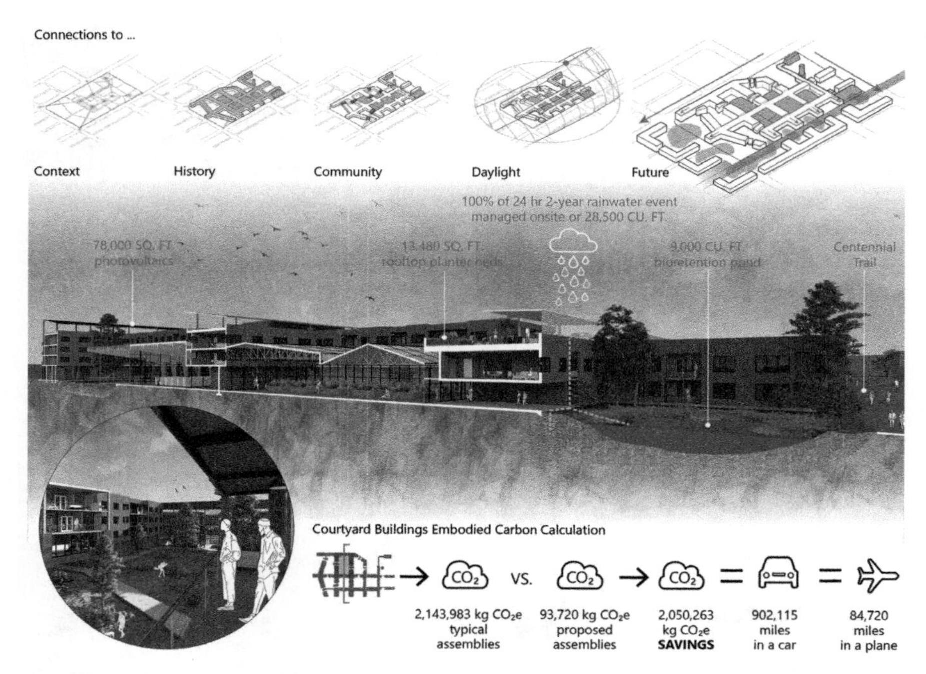

FIGURE 12.7 Example project that inverts the existing structures from indoor to outdoor spaces. Illustrations: north–south section, diagrams, render, on-site water management calculations and embodied carbon calculations for one residential courtyard.

Source: Jaclyn Allen, Amy Borer and Sydney Troy

structures with new programs that increase in height from south to north in response to solar angles. This results in several courtyards that are interconnected by a series of pathways between the major nodes on site. The intersection of these pathways with the new perimeter programs establishes the community nodes where multi-purpose communal spaces are located. The master plan in Figure 12.8 is designed by thinking beyond the boundary of the site and creating a "Green Belt" parallel to the Centennial Trail between Falls Park to the west and Black Bay Park to the east. This Green Belt adds over 200 trees on site and over 800 trees in the area between the two parks. The trees planted on this site can mitigate the amount of carbon produced when every resident in the Post Falls area drives nearly 650 miles for an entire year.

Phase Four: Final Design and Presentation takes place for the remainder of the semester where student teams address comments from phase three and continue to detail the design. They develop their final presentation material for the course in alignment with the AIA Committee on the Environment Student Competition. At the end of the semester, students present to faculty and external reviewers in digital slides, two presentation boards, one physical model (1/128″ = 1′ – 0″) and a 1500 word written narrative.

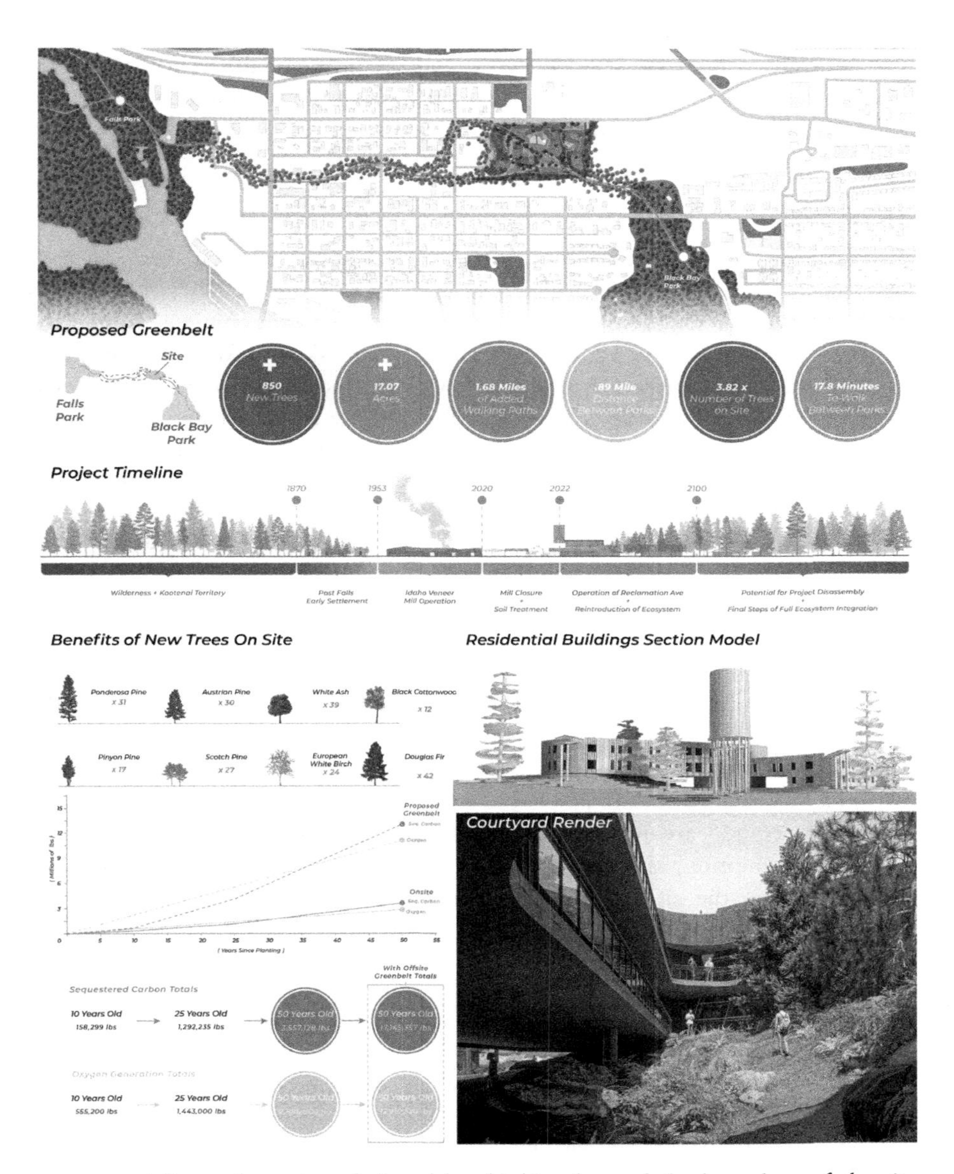

FIGURE 12.8 Example project designed by thinking beyond the boundary of the site and connecting Falls Park to Black Bay Park through a Green Belt. Illustrations: master plan, timeline, diagrams, physical model and render.

Source: Karly Ennis, Jacob Lewis and Ryan Quinn

The instructor reviews student work and provides comments to each team. Student teams apply comments to their presentation and submit their entry to the competition at the beginning of the Spring semester.

Assignments

Discussions, design presentations and peer evaluations are the three primary assessment methods used to evaluate students in this studio. Discussion boards set up in Canvas are used to form responses and generate opinions around the course required readings. Additionally, students are asked to form questions that become the basis for the Q&A sessions with the industry experts. Design deliverables are formally presented verbally and visually to faculty and professional practitioners at the end of each phase of the semester. Peer evaluations occur twice throughout the semester to gather input on team members responsibilities and performance, and to address any challenges associated with team dynamics.

Analytics

Students are introduced to software tools early in the semester so they can iteratively develop their designs and quantify the impact of their decisions on the natural environment as well as assess building performance. Due to the large scale of the project, students perform simulations on a critical portion of their project which helps reinforce their vision for the site. Tally for Revit[5] is introduced to help students assess the embodied carbon. Solemma ClimateStudio[6] and Sefaira[7] are introduced to help students assess annual operational energy consumption, and annual indoor daylight quality. Climate Consultant[8] is used during the site analysis phase to study the climate and identify the optimum design strategies for the site.

Required Readings

Association of Collegiate Schools of Architecture. "Program." 2024 COTE Competition, Accessed August 21, 2023. www.acsa-arch.org/competitions/2024-cote-competition/program/.

Bollack, Françoise Astorg. *Old Buildings, New Forms: New Directions in Architectural Transformations*. New York: Monacelli Press, 2013.

City of Post Falls. "City of Post Falls Comprehensive Plan." Community Development Online. Accessed December 3, 2023. www.postfalls.gov/PZDept/pzforms/Planning/CompPlan.pdf.

City of Post Falls. "City of Post Falls Smartcode." Community Development Online. Accessed December 3, 2023. www.postfalls.gov/PZDept/SmartCode/SmartCode.pdf.

City of Post Falls. "City of Post Falls Downtown Public Space Feasibility Study 2020." Parks and Recreation Planning Documents & Studies. Accessed December 3, 2023. www.postfalls.gov/PZDept/SmartCode/SmartCode.pdf.

City of Post Falls. "City of Post Falls Community Center Feasibility Study." Parks and Recreation Planning Documents & Studies. Accessed December 3, 2023. www.postfalls.gov/PZDept/SmartCode/SmartCode.pdf.

LMN Architects. "Path to Zero Carbon Series." LMN Research. Accessed August 24, 2023. https://lmnarchitects.com/lmn-research/path-to-zero-carbon-series. Framing the Challenge (1–3); Fundamentals (6, 7); Circular Economy and Existing Buildings Reuse (9–10); Structure, Envelopes and MEP (11, 12, 14).

Robiglio, Matteo. RE-USA: 20 *American Stories of Adaptive Reuse, a Toolkit for Post-Industrial Cities*. Berlin: Jovis Verlag GmbH, 2017.

Speck, Jeff. *Walkable City Rules: 101 Steps to Making Better Places*. Washington, DC: Island Press, 2018.

Recommended Readings

Knowles, Ralph L. *Sun Rhythm Form*. Cambridge, MA: MIT Press, 1981.

LMN Architects. "Path to Zero Carbon Series." LMN Research, August 24, 2023. https://lmnarchitects.com/lmn-research/path-to-zero-carbon-series. Fundamentals (4, 5, 8); Operational Carbon (13).

Outreach Opportunities

The key collaborators from professional practice during the 2023 version of this studio are A&A Construction who are the current developers of the site and provided access to the site as well as their proposed future development plans. LMN Architects provided students with the technical knowledge on carbon neutrality, workshops on building performance analysis tools and feedback to students on their projects during formal reviews as well as through online communication outside of the official class time.

Course Assessment

National Architecture Accreditation Board Accreditation Criteria

The National Architecture Accreditation Board (NAAB) student criteria covered in this studio are:

1. Health, Safety and Welfare in the Built Environment;
2. Regulatory Context;
3. Technical Knowledge; and
4. Design Synthesis.

Additionally, the NAAB program criteria covered in this studio are:

1. Design;
2. Ecological Knowledge and Responsibility; and
3. Research and Innovation.

Lessons Learned

Over the past three years of teaching this studio, several iterations have been tested as it pertains to defining the starting point for designing the project. In the first version of the course from 2021, a deductive approach was utilized whereby students began with designing the overall master plan in phase two, then zoomed into the design of the existing structures in phase three. A reverse inductive approach was tested in version two of the course from 2022 whereby students began with designing the existing buildings in phase two, then zoomed out to the design of the overall master plan and its connection to the surrounding context in phase three.

It was found that the inductive approach resulted in preferable outcomes as beginning with the existing structures provided students with design inspiration and the structural grids occurring on site provide guidelines for organizing the master plan as well as the dimensions and spatial qualities of the new programs being introduced. In the third and latest version of the studio the inductive approach was followed, and teams were given the flexibility to alternate between designing either inductively or deductively in phase two in coordination with the instructor.

Looking Ahead

The introduction of articles from the Path to Zero Carbon Series as required readings for the first time in version three of the studio has had a highly positive impact on students' level of awareness and understanding of carbon neutrality. This was confirmed by their in-depth responses to the readings through discussion boards posted in the learning management system coupled with Q&A sessions with the authors from LMN Architects. This will continue to be implemented in the studio moving forward.

Building performance analysis using Tally was introduced for the first time in the third version of the studio as well. This has been useful in providing students with direct feedback on their enclosure assemblies materials decision-making in the 3D software they use to visualize the design. Future versions of this course will continue to refine the workshops related to this software and make it a required deliverable earlier in the semester beginning from phase two.

Notes

1 Elefante, Carl. "The Greenest Building … Is the One That Is Already Built." Insights. Accessed December 2, 2023. https://carlelefante.com/insights/the-greenest-building-is/
2 Jacobs, Jane. *The Death and Life of Great American Cities*. New York: Random House, 1961.
3 Russo, Michele. "44 Percent of Design Activity Is Devoted to Existing Building Projects …" American Institute of Architects: Knowledge. Accessed November 1, 2018. www.architectmagazine.com/aia-architect/aiaknowledge/44-percent-of-design-activity-is-devoted-to-existing-building-projects_o

4 Logan, Katharine. "Renovate, Retrofit, Reuse: Uncovering the Hidden Value in America's Existing Building Stock." American Institute of Architects, 2019. Accessed December 2, 2023. https://content.aia.org/sites/default/files/2019-07/RES19_227853_Retrofitting_Existing_Buildings_Report_Guide_V3.pdf
5 Building Transparency. Computer software. *Tally*. Accessed December 2, 2023. https://choosetally.com/download/
6 Solemma LLC. "Climatestudio." Accessed December 2, 2023. www.solemma.com/climatestudio
7 Trimble. "Sefaira." Accessed December 2, 2023. www.sketchup.com/products/sefaira
8 Society of Building Science Educators. "Climate Consultant." Accessed December 2, 2023. www.sbse.org/resources/climate-consultant

13

THE HA/F RESEARCH STUDIO & SEMINAR

Kelly Alvarez Doran

Course Overview

Course Name

The Ha/f Research Studio & Seminar

University / Location

University of Toronto / Toronto, Ontario, Canada

Targeted Student Level

- Graduate
  - 2nd or 3rd year, Master of Architecture, 3-year program

Course Learning Objectives

In this course, students will:

- Learn Life Cycle Assessment (LCA) theory, methodology and practice.
- Undertake case studies of existing buildings to understand both the embodied and operational impacts of buildings with a focus on carbon emissions.
- Work together to develop comparative analysis between buildings in the study set to uncover the key sources and design drivers of emissions in building design and construction.

DOI: 10.4324/9781032692562-17

Course Learning Outcomes

Successful completion of this course will enable students to:

- Through case study research, engage and interview real world projects and the practitioners who designed and oversaw their construction – exposing them to the depth and range of considerations and constraints architects work with through project design.
- Lead participating practitioners – the Ha/f Studio is often their first exposure to Life Cycle Assessment and considerations of embodied carbon in their own designs. Many of the participating practices have since implemented LCA into their practice, and hired graduates of the Ha/f Research Studio to support their efforts.
- Gain a detailed understanding of construction documents, specifications, LCA, material provenance and operational performance of contemporary construction.
- Comprehend the holistic assessment and analytical skills required of our industry to both identify and find alternative solutions to help reduce the carbon footprint of buildings by half, this decade.

Course Methodology

Philosophy

The Ha/f Research Studio was established in 2020 to answer the basic question posed to the world by the Paris Agreement: "How do we halve the greenhouse gas emissions of the built environment this decade?" The initial intention of the studio was to catalyze a conversation around embodied carbon in Toronto, and through it to expose students, practices, building owners and policy-makers to the methods available to account for a building's life cycle emissions.

To answer this question, we must first answer "half of what?" – we need to establish benchmarks to track future progress against. Given that buildings represent roughly 40% of annual global emissions globally, with a quarter of that attributed to the upfront, embodied carbon emissions of new construction, there is an urgency and imperative for architects to acquire the knowledge and skills necessary to quantity and improve upon our out-sized emissions. Moreover, as our buildings continue to reduce their operational footprints and as our grids decarbonize, the majority of a building's life cycle emissions are those of its construction – the "cradle to gate" upfront embodied carbon emissions associated with the extraction, harvesting, processing, manufacturing, transportation and construction of the building's materials.

This has been our industry's blind spot for decades – more so in academia where the vast majority of our training and course offerings have focused solely on operational energy consumption alone. Through a case-study approach, the

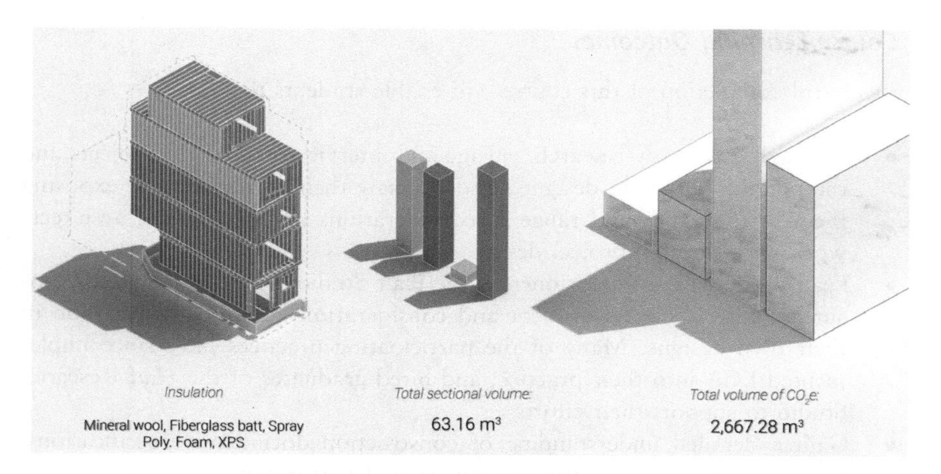

FIGURE 13.1 Representation of a project's insulation systems, their relative material volumes and the relative embodied carbon emissions of those materials.

Ha/f Research Studio serves as a regionally replicable template for a heuristic approach to empower students and practitioners with the tools and data required to take immediate climate action and to advocate for the necessary changes in our industry and our academic offerings. The first two years of the research studio looked at multi-unit housing in Toronto and mass timber buildings across North America and Europe. For the past two years, in support of the University's Climate Positive plans the course has worked with the university to assess its own building's embodied and operational carbon emissions. The Studio's results have been published immediately in trade publications to broadcast the knowledge generated in the hopes of influencing practitioners and policy makers.[1]

As a means of direct advocacy, the data and insights generated in the first Ha/f Studio focused on Toronto's multi-unit housing was presented to the City of Toronto and has subsequently led to the co-authoring of an embodied carbon policy. The policy was adopted by Council in the spring of 2023 – making Toronto the first city in North America to set maximum embodied carbon targets for new construction. In just three years the Ha/f Research Studio has revealed both the need and opportunity for architectural education to shape and inform the changes required of our practices, bylaws, codes and regulations to enable the radical reductions required of the built environment.

Definitions

Carbon neutrality – Architecture is never neutral. A building is inherently emissive – both in its construction and operations. Any building claiming to be net-zero or climate positive in and of itself is not accounting for the entirety of its making and maintenance. Buildings are the aggregation of material, energy, carbon and labor flows that extend well beyond their sites. The challenge is understanding those flows and where architecture can intervene within them to maximize socio-economic impacts and minimize environmental impacts. For architecture to truly move towards a state of climate positivity it must first work to fully comprehend these impacts and re-route its investments into regionally abundant sources of material and labor. Buildings that move closer to their tributaries of supply are inherently more connected to their impacts, and more able to make more informed, more holistic choices. Pedagogically, this requires us to equip architects to engage and understand their terroir. Buildings of the earth, or the fields and forests, and of the place can move us towards climate positivity. We need to train a generation of urban miners and bio-based innovators. We need to redefine our metrics. We need to leave terms like "neutral," "zero" and "sustainability" behind us as they are no longer useless or ambitious enough for the challenges we face. "Positive," "less" and "sufficiency" could be useful terms to define together.

Embodied carbon – The volume of carbon dioxide equivalent emissions associated with the making, use, disposal or reuse of material or product

Life cycle assessment – A tool to assess the potential range of environmental impacts a building, landscape or infrastructure could have over the course of its functional life.

Operational carbon – The volume of carbon dioxide equivalent emissions associated with the heating, cooling, ventilation and plug loads of a building, landscape or infrastructure over the course of its functional life.

Whole life carbon – The total volume of carbon dioxide equivalent emissions associated with a building's life cycle construction, operations, disposal or deconstruction.

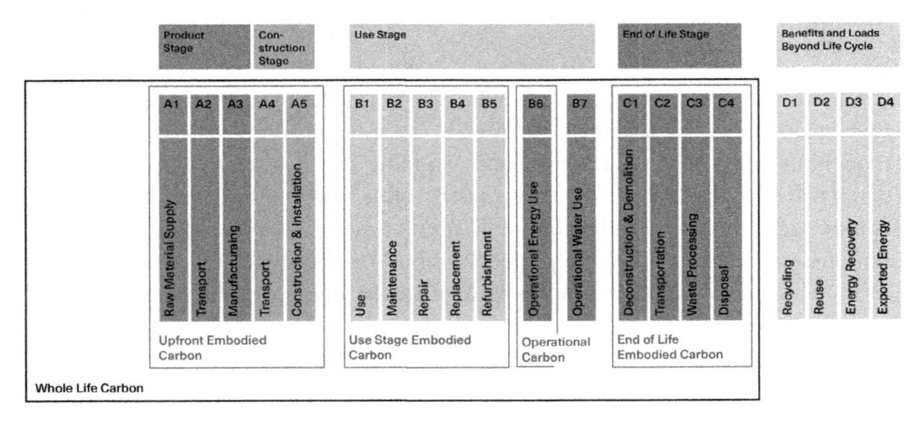

FIGURE 13.2 Illustrated table of Life Cycle Assessment Phases per EN 15978 and ISO 21930.

Implementation

General

To catalyze regional and typological embodied carbon benchmarking the studio engages practitioners through a study set of recently completed existing buildings that represent the breadth of contemporary construction practices. Through these case studies – an approach typical of law and business schools – students are tasked with developing a deeper understanding of a project's construction drawings and its specification before selecting a representative section of the project to develop a detailed digital model from.

For many students this is also their first exposure to construction documentation. Additionally, students are required to conduct interviews with the design teams to establish an understanding of the key drivers of design decision making; ranging from zoning and code requirements, to client requirements and preferences, to cost and value engineering, and to an office's own preferences and typical details. Students then conduct comprehensive Life Cycle Assessments of the projects and put together an illustrated report of their findings and then collaborate as a class to establish a comparative analysis of the broader study-set.

Assignments

The methodology of the Ha/f Research Studio is designed to enable students to complete a detailed whole life carbon assessment over the course of a thirteen-week term. The course – initially structured as a research studio – is now a research seminar co-taught with Assistant Professor Alstan Jakubiec and is structured in three modules:

Module 1 (Weeks 1–6): Embodied Carbon Life Cycle Assessment

Working in groups of two, students are assigned a project and given access to a complete set of construction documents in both PDF and 2-D digital drawing format. Practices are asked to provide 2D drawing sets in place of Revit or BIM models in order to have the students assemble the fragments of plan, section, details and specification into a complete understanding of how the building comes together. In consultation with the instructor – students select a representative structural bay of the project and build a detailed 3D model of that bay as a means to extrapolate a Bill of Quantities of materials of that model.

Students then conduct hour-long interviews of participating practices to ask questions of "why" and "how" of the building's design and construction. These interviews often reveal unexpected design drivers and ask unforeseen questions of practice that further reveal either blindspots or deep bias in current construction paradigms. Finally, students use the LCA tool One-Click LCA to associate the quantitative information of the model with the Environmental Product Declarations of each product as specified in the construction documents. The outputs of the LCA focus on the Global Warming Potential (GWP) of each project as expressed in kgCO2e (kilograms of carbon dioxide equivalent) through an illustrated report in the form of a slide presentation.

Module 2 (Weeks 7–11): Operational Assessment

Groups then work to compare modeled vs. actual energy consumption across the study set. Working with ClimateStudio and EnergyPlus students establish an understanding of modeled energy consumption based on the occupancy, thermal performance and mechanical systems employed in the buildings. Groups then match up the modeled energy consumption with the actual consumption data provided by the building owner – in the initial case being the University of Toronto – to assess how alignment and to understand where divergences occur. Field trips to study buildings are conducted over the course of this module with building operators for students to become familiar with the various building and mechanical systems employed.

Module 3 (Weeks 12–13): Whole Life Carbon & Comparative Analysis

Students merge the LCA and energy model and usage data to create a whole life carbon assessment of the study projects to visualize the proportion and rate of each project's global warming potential.

Finally – working as a studio – students compare their findings to develop comparative assessments between the family of projects to better understand broader trends and how specific design decisions in each project relate to larger factors like typology, building code, zoning and bylaws, and use. This comparative study has led to several unexpected discoveries; over the four years of the study comparative studies have revealed the following unexpected observations of the study-sets:

1. **Our buildings are icebergs.** Foundation works, underground parking structures, and below-grade floor area have disproportionate impacts on a project's embodied carbon. For mid-rise and high-rise structures, between 20 to 50% of each project's total volume of concrete was below grade.
2. **The impact of parking.** The spacing requirements of parking produces a structural grid that is often out of alignment with that of a typical residential floor plate. Projects that employ transfer slabs or structures to mediate these alignments have higher embodied carbon as a result.
3. **Bio-based materials are lower carbon.** A comparison of embodied carbon and thermal performance of building envelopes revealed that the highest performing, lowest carbon facades were those that utilized a range of bio-based materials (mass timber, wood fiber board, cellulose and wood siding). Moreover, when taking into account biogenic sequestration these facades could be deemed climate positive assemblies.

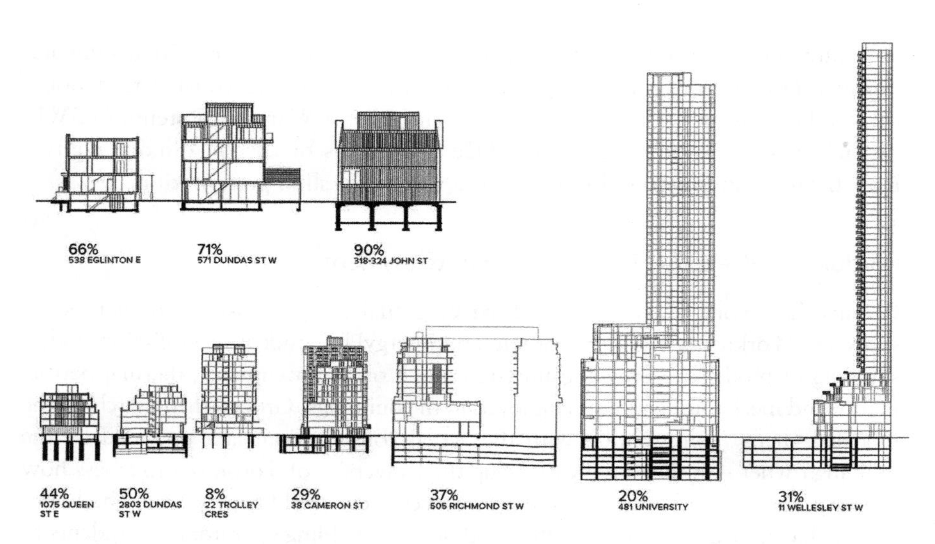

FIGURE 13.3 Percentage of embodied carbon below grade in Toronto multi-unit residential buildings.

Analytics

LCA and energy modeling are inherently quantitative analytical tools that can serve to not only illustrate the impact of a design decision, but also identify alternative options to mitigate environmental and social impacts.

To perform the LCA's students rely on either Revit or Rhino to create the models and extrapolate the quantitative information from those tools into spreadsheets. Students then use LCA tools One-Click LCA and EC3 to associate the quantities with the specific environmental product declarations of each material provided by specific manufacturers or industry representative bodies. The LCA analysis is ultimately a door that opens a series of other doors revolving around the material provenance of construction – where is it made? how is it made? and who made it? is the line of enquiry that students are encouraged to take.

Energy modeling analysis is likewise inherently analytical. By comparing the actual utility data with their models of the buildings, students have revealed the disconnections within our contemporary definitions of "efficiency" and "sustainability." The broad trend across the 40 buildings studied in the Ha/f Research Studio is that older buildings use less energy operationally, and have lower carbon footprints to construct. This analysis throws every assumption into question, and underlines the importance of educating and enacting whole life carbon analysis as part of the education and practice of architects.

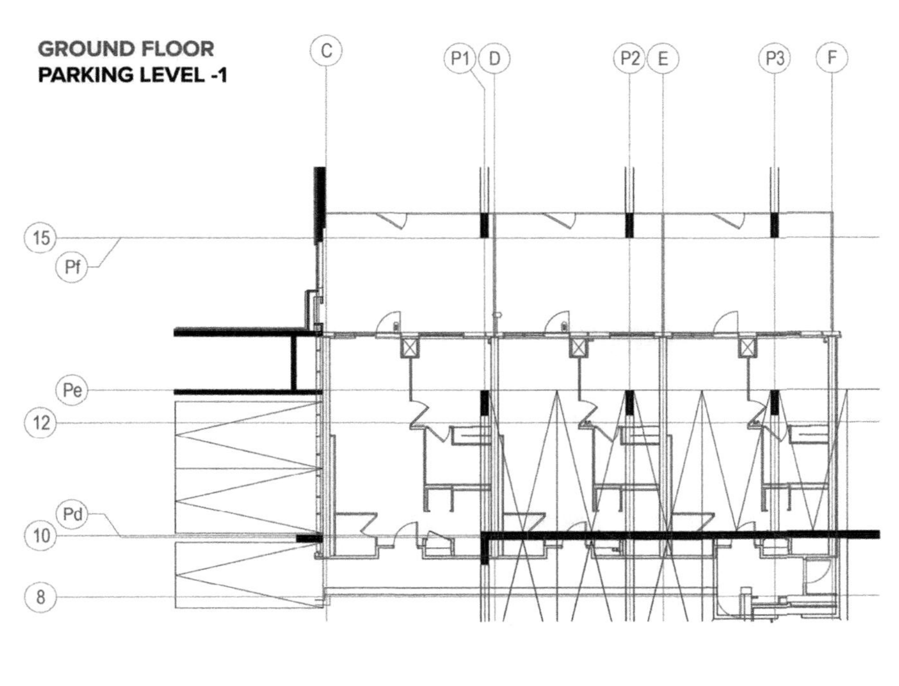

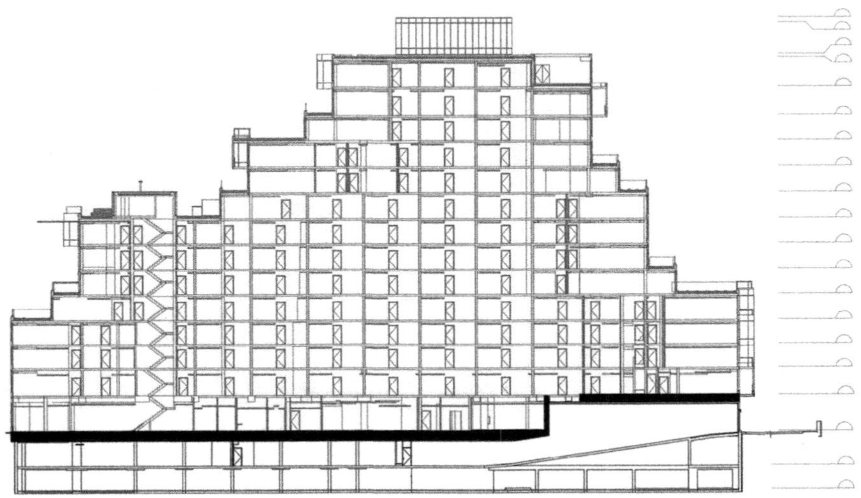

FIGURE 13.4 Left: An overlay of the parking level and ground floor residential structural grids shows the disjunction between the two systems—a common issue in multi-unit residential structures with underground parking. Right: A 1 m-thick transfer slab was needed to negotiate between the two structural grids—an element that represents 14% of the total project concrete.

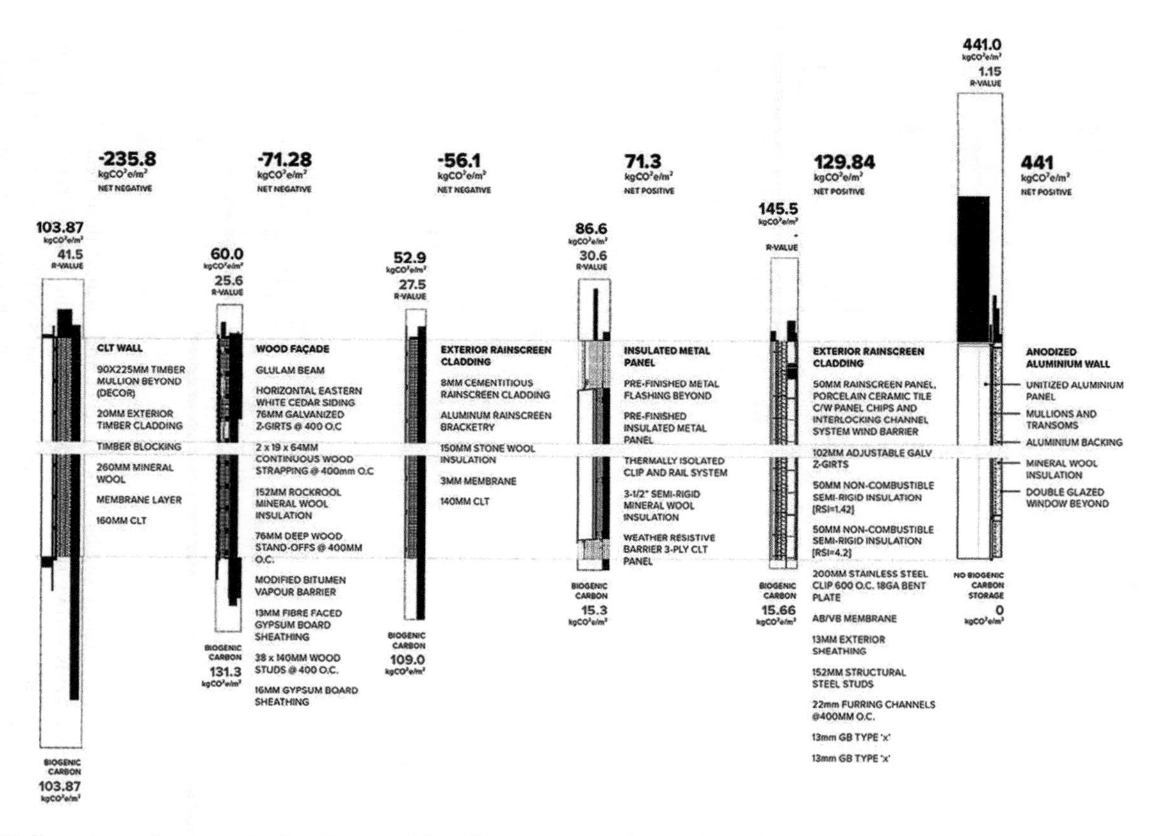

FIGURE 13.5 Wall sections of case study facade assemblies illustrating R-value, embodied carbon and biogenic sequestration.

Required Readings

Alvarez Doran, Kelly. "Thinking Through Material Flows." *Architects Newspaper*, 2023. www.archpaper.com/2023/05/rather-primary-focus-energy-efficiency-architects-should-consider-more-holistic-assessments-reduce-carbon-emissions-construction/
Gibbons, O. P and Orr, J. J. "How to Calculate Embodied Carbon" (Second Edition). Institute of Structural Engineers, 2022. www.istructe.org/resources/guidance/how-to-calculate-embodied-carbon/
London Energy Transformation Initiative. "Embodied Carbon Primer." 2020. www.leti.uk/ecp
Moe, Kiel. "Metabolic Rift, Gift, and Shift." eflux 2020. www.e-flux.com/architecture/accumulation/345105/metabolic-rift-gift-and-shift/
Simonen, Kathrina. *Life Cycle Assessment*. Routledge, London, 2014.

Outreach Opportunities

The Ha/f Studio was founded to catalyze conversations in Toronto, Ontario, and beyond around embodied carbon. Upon completion of the first cohort's analysis of multi-unit residential buildings in Toronto the findings were shared with the City of Toronto's planning team with the intention of initiating a discussion about how the City could and should regulate embodied carbon in line with their policies surrounding operational carbon via the Toronto Green Standards. That conversation led towards jointly securing a grant to expand the data set to serve as a benchmark to peg future policy against. With me, Ha/f Studio graduates worked with the City to collect LCA data from the broader industry and co-lead stakeholder engagement sessions to shape a policy that sets maximum embodied carbon targets that ratchet down over the coming decade in line with the existing Green Standards tiered system. The policy co-authored by the Ha/f Studio was adopted by City Council in the spring of 2023.[2]

Each year's work has been published nationally via *Canadian Architect*[3] to share the findings nationally. This exposure has led to several speaking engagements over the past three years with offices, professional associations, conferences and panels. Graduates of the Ha/f Studio have gone into offices and become key members of the respective sustainability teams – further accelerating the transition to whole life carbon approaches across the profession.

Finally – the work of the studio has led to the formation of a company between myself and two graduates of the inaugural studio. Ha/f Climate Design was founded in 2022 to work with architects, engineers, planners and city builders to train and advocate for the acceleration of whole life approaches in policy and practice. Ha/f's co-founders teach at the University of Toronto, the University of Waterloo, and are now working with the Government of Canada and the Royal Architectural Institute of Canada to provide training to over 1,500 architects in 2024.

Course Assessment

Student response to the course has been very encouraging. On numerous occasions students have expressed their satisfaction with the case study approach as a means of being introduced to the details and realities of architectural practice that they'd not encountered prior in either school or practice experience. Their interactions with leading practitioners have helped them establish connections with industry that has led to immediate employment opportunities, speaking opportunities, and jointly working together on future grant applications.

Looking Ahead

This year's Ha/f Studio concludes a two-year focus on the University of Toronto's own buildings – the results of which will be shared in the fall of 2024. Ha/f will be offering training workshops to students, teachers and practitioners in Scotland, Australia and across Canada to enable other institutions to replicate the case-study approach. We thank the editor and publisher for the opportunity to share the pedagogy and impacts of the Ha/f Studio to date and look forward to seeing it replicated elsewhere.

Notes

1 See www.canadianarchitect.com/why-we-need-embodied-carbon-benchmarks-and-targets-in-building-standards-and-policies-an-open-letter/.
2 See https://halfclimatedesign.com/research-policy.
3 See www.canadianarchitect.com/technical-mass-timber-through-a-life-cycle-lens/.

14

BOXES AND DOXA

Learning from the Solar Decathlon Design Challenge

Jonathan Bean

Course Overview

Course Name

Design Competition; Solar Decathlon Design Challenge Option Studio

University / Location

The University of Arizona / Tucson, Arizona, USA

Targeted Student Level

- Undergraduate
  - 4th year, Bachelor of Architecture, 5-year program (Option Studio Course)
  - 2nd, 3rd, or 4th year, any other program (Design Competition Course)
- Graduate
  - Any year or program (Design Competition Course)

Course Learning Objectives

In this course, students will:

- Gain a deep respect and curiosity for the field of building science, grounded in a knowledge of the basic principles that govern buildings' energy use, durability and occupant health.

DOI: 10.4324/9781032692562-18

- Build experience with iterative problem-solving, increased creativity, resilience and the opportunity to learn new skills quickly (four of today's employers' most-wanted skills.) As we'll be working in teams with tight deadlines, there will be plenty of opportunity to improve your abilities in project and time management.
- Polish verbal and graphic communication skills, with a special emphasis on negotiating the difference between the specialized language of architecture and design culture and effective communication with a broader public.
- Create meaningful content for your portfolio to assist in securing future employment opportunities.

Course Learning Outcomes

Successful completion of this course will enable students to:

- Design a building to meet the requirements of the Solar Decathlon Design Challenge competition. The design and submission will meet the requirements for the selected building type as defined by the Solar Decathlon Design Challenge Rules document and the relevant Phius ZERO criteria.
- Use dynamic hourly energy modeling to inform decisions at all stages of the design process.
- Develop an area of expertise that corresponds to at least one of the ten contest criteria defined by the competition. (These categories shift slightly yearly; in the past the list has included architecture, engineering, market analysis, durability and resilience, embodied environmental impact, integrated performance, occupant experience, comfort and environmental quality, energy performance and presentation.)

Course Methodology

Philosophy

When teaching any course, I aim to help each student build a greater sense of agency and direction. My colleagues are sometimes surprised by what they perceive as a disjunction between the open-ended goal of building student agency and the rigid requirements of the Solar Decathlon Design Challenge competition. For example, students must quantify the performance of their designs using several specialized tools, such as the software program REM/Rate, and they are required to estimate costs using RSMeans data. From some perspectives, these requirements appear to be forcing students into a box – or, worse, to design one!

Busting out of a box, of course, is the implicit lesson of so much of what we teach in architecture school. In the early years of architectural education, we teach students a new verbal and spatial vocabulary: planes and points, solids and voids, axis and terminus. The pedagogical intent is to rewire student's brains and help them to see and think like an architect, able to appreciate the subtlety of a James Turell skyspace, and to see through the ornament and view classical architecture as an exemplar of proportion and order. This shift in ways of seeing and thinking about space is central to the identity of an architect. Students often come to the competition course intimidated by the need to comply with long lists of requirements, to work in teams and to run calculations. The approach I take helps students explore the different pathways they can take in architecture. I start by reframing a core question: asking the students to quantify what could be done with the energy the efficient building they will design is *not* using. This inversion of the usual approach to saving energy intentionally challenges the students to question assumptions. This move and the course pedagogy are informed by two influential ideas from sociology, which are introduced in the following section.

Definitions

Carbon neutrality – I think the best we can do is acknowledge that the definition of this term is shifting, and will probably continue to do so. The concept of carbon neutrality stands in strong opposition to the idea of *doxa*. Rather than being unseen and unquestioned, it is the subject of active debate. Any answer to the question depends on a number of assumptions, starting with the time horizon for the calculation. As such, carbon neutrality is fundamentally and philosophically a question of accounting. I expand on this below.

Doxa – The unspoken and unseen, or tacit, rules that undergird cultural norms in a specific field. Cécile Deer describes *doxa* as "refers to pre-reflexive, shared but unquestioned opinions and perceptions conveyed within and by" a field of practice.[1] *Doxa* are present in religious traditions, where differences of opinion on some topics may be acceptable, while divergence on others can constitute grounds for exclusion. *Doxa* is often difficult to articulate. It can be easier to identify what is ruled out: in the case of architecture, mere building. Members of a field – in this case, professional architects, architecture students, professors of architecture and others with a valid claim to membership – question *doxa* in public at their own peril. Efforts to encourage interdisciplinary education, for example between architects and engineers, often bump into *doxa* without naming it as such.[2]

Field and doxa – The sociologist Pierre Bourdieu refers to a field of practice as a relational entity. Although field is an abstract concept, in some ways a field is akin to a football field. Players (architects and aspiring architects) occupy a position on the field that exists in relationship to others who are also in the field. To play the game, all must understand the ground rules. Fields also exist in relationship to other fields; for example, the field of architecture can be understood to exist in the context of the building industry alongside engineering, real estate and construction. But within the field itself, architecture functions as an autonomous, creative and expressive endeavor with rules similar to other cultural fields such as music and fine art. This insider view of architecture distinguishes between the dominant producers or avant-gardes, which indicates architects such as Frank Gehry or Zaha Hadid, and subordinate producers, which includes most everyone else.[3]

Implementation

Learning to See Boundaries, Identify Adjacent Fields and Surface Doxa

Most of the students I have led through the Design Challenge competition are in the fourth year of a five-year architecture program. At this point, they are well versed in the specialized professional language of architecture, and they are becoming increasingly aware of the *doxa* that underpin success in the field. For example, students learn that they are taken more seriously when they use the term "fenestration" rather than "windows," or when they adopt the convention of talking about the design of the building as if it generated itself. For example, a student might present a design by saying that the elevator core "wanted" to be near the mechanical chase and egress route. For those immersed in the field, this choice of language indicates that the designer has developed abstract rules to align practical and aesthetic requirements by coordinating the demands of vertical circulation, space conditioning and ventilation and fire safety. But for those outside of architecture, this can be confusing: elevator cores don't have agency of their own, so how can they "want" anything?

In the context of an architecture studio, using the right terminology is a high-stakes game. In many architecture programs, students are seen as legitimate only if they follow the rules and adhere to doxa. In turn, grades, course placements and recommendations are, to some degree, a reflection of how deeply a student has internalized the doxa and how fluent they have become in the specialized language of the field.[4] I stress here that it is *not* the specialized language that is the problem. Most people would not want cancer researchers, pharmacists, or air traffic controllers to be forced to speak only in simple terms. At the same time, many of us appreciate the effort taken to translate specialized terminology into concepts that non-specialists can understand. From my perspective, the field of architecture has done well with building up specialized terms, but could do more to translate the value of what we do to other fields.[5]

Several aspects of the Design Competition require students to burnish this skill, but none more so than the juried competition. Projects are judged by a panel of four to five jurors. In my experience with the competition, students are lucky if *one* juror is an architect. Others on the panel might be mechanical engineers; building code officials or consultants; representatives of building equipment manufacturers; or members of industry organizations with a distinct point of view on, say, plastic foam insulation. Because a question-and-answer session is included in the scored evaluation, students may be peppered with questions that would not come up in a typical studio review: what was behind

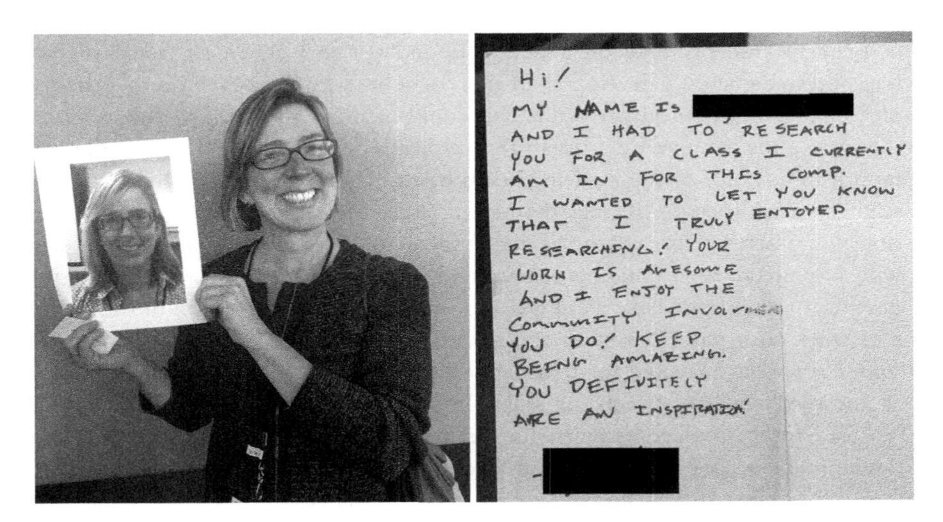

FIGURE 14.1 Juror Diana Fisler with a thank you note that a student wrote her.

the specification of a particular brand or size of mechanical equipment? Why did you exceed code minimum levels of window performance? Why didn't you use any foam insulation? The Q&A can be a minefield for students accustomed to shaping their language to the norms of the field of architecture.

FIGURE 14.2 The Attached Housing jury for the 2022 competition included a researcher and biomimicry expert; a passive house architect; a housing developer; and a civil engineer.

To help students take on a new role and see themselves as translators of ideas between fields, I developed an exercise where students take on the persona of a Solar Decathlon juror. While this could be done generically using a list of job titles, the competition organizers typically announce the jurors about a month in advance of the final in-person event. This makes it possible to assign students a real person. Rather than being a generic representative of an insulation manufacturer, for example, one student took on the role of Diana Fisler, who was at the time Building Science and Innovation Leader for the insulation manufacturer Johns Mansville. I give students a printed photograph of the juror's face stuck to a paint stir stick and ask them to come to the next class with a resume they have created for that person, along with a short-written piece, such as an op-ed, series of tweets, or LinkedIn posts totaling around 250–500 original words written in their assigned juror's voice. The idea is to get the students thinking and seeing from a perspective outside the field of architecture.

As the student teams finalize their designs in the two to three weeks leading up to the in-person event, they practice their presentations repeatedly to their jurors, who play the part complete with name tags and life-sized photographs. Every time I use this role playing method, which was inspired by Mark Carnes's book *Minds on Fire*,[6] I see more benefits. First, students who otherwise shy away from confrontation will ask tough questions – and even take on independent research on their own time to anticipate their juror's questions. On several occasions, students have correctly anticipated either the general outline or the specific content of actual juror questions, which sets up the teams for a home run answer. Second, it's just fun. Carnes explains at length how part of the fun of any game is transgressing rules. With the tradition of tutelage weighing heavily on architectural education, allowing students to try on the role of an expert is liberating. Third, the introduction of the juror role is akin to doubling the number of people in the classroom. Not only do the students have to consider a multitude of perspectives that they might not otherwise be familiar with, the students also have to flip back and forth between their newly assigned role and their former and ongoing role as a member of a competition design team. This helps students understand the adage that to be an expert, you must know both what you know and what you don't know. In short, the exercise helps them to see how *what* knowledge counts and *who* values that knowledge is specific to a field.[7]

Analytics: What Are We Counting?

In general, students have a vague idea of what carbon neutrality is; and most seem to perceive it as a good or desirable trait for buildings and their designers to pursue. To help students think through why carbon neutrality, and related terms such as net-zero and climate-positive, are hotly debated and difficult to define, I have found it helpful to start with an example borrowed from Prof. Cindy Guthrie, who teaches accounting at Bucknell University.

Prof. Guthrie related to me how students often walk into her class either overconfident – accounting is just math, right? – or borderline terrified – because accounting is math! To help them see the challenges of accounting from a different perspective, she developed a brilliantly simple exercise. Teams of students each get a box. Each box contains many boxes of candy that the students are to inventory. Inventory, of course, is a routine and visible task that's familiar to most students, even if their only exposure has been venturing into a store during the disruption of counting. The students quickly discover complications. Some of the boxes of candy are expired; others will expire in the coming days. A few boxes are open or damaged. Some contain the wrong kind of candy; others are filled with rocks or paperclips. What, exactly, counts? Candy that will be saleable in the next week, month, or year? Whole boxes or pieces? Whose job is it to fix the supply chain or quality problems that are allowing a substantial amount of non-candy into the warehouse in the first place?

The candy exercise parallels core questions around carbon and climate accounting. First: what are we counting: carbon dioxide or global warming potential? Most students are well aware that carbon from the construction and operation of buildings is a key concern in global warming. However, Project Drawdown identifies better refrigerant management as the number one lever for reducing global warming.[8] This means eliminating, as much as possible, leaks from the refrigerant used in air conditioners and heat pumps and large-scale cooling systems, such as those in grocery stores and cold storage warehouses, and using less refrigerant in the first place, which can be accomplished with highly efficient buildings and equipment. Perhaps there are other kinds of candy that we *should* be counting! For example, when methane leaks directly to the atmosphere without being ignited, it has a relatively high GWP-100 of about 27, meaning it is 27 times worse than an equivalent amount of carbon dioxide over a 100-year timespan. Considering how methane is delivered to buildings provides a route to discuss the systems issues akin to the broken candy supply chain. Of course, combustion of methane in buildings, which is typically done to heat spaces for people or to fuel industrial processes, creates a direct plume of carbon dioxide emission. But what about the significant emissions from leaks in production, storage and transportation pipelines? How should this baseline level of emissions be accounted for? What happens to these emissions levels in a future where we are both using less methane, but where methane is a critical fuel because it is the main source of power for backup generators and peaker plants? What is the responsibility of the building designer to consider these systems-level questions in the design of an individual building?

We also discuss how energy is calculated and quantified. While the competition requires or encourages student teams to use specific software packages, such as Ekotrope, I have required my students to develop dynamic hourly energy models of their designs. To do this, we use the dynamic function of WUFI Passive, which is essentially a static energy modeling package that sits on top of the dynamic functions of WUFI Plus. We use WUFI because of path dependency – it is a dynamic energy modeler that I know and can support – but any other program that can produce and export an hourly model of energy use would support one of the core learning outcomes, which is to use dynamic hourly energy modeling to inform architectural design.

Static measurements such as energy use intensity (EUI) and modeling tools such as Sefaira that present users with monthly or annual averages of energy use are helpful entry points into the broader world of building performance modeling. To understand how buildings may work in the future, for example as Grid-Interactive Efficient Buildings, or GEBs, students need to understand how thermodynamic flows vary with time, and how those flows shift daily and seasonally. This helps students understand how the advantages of energy efficient buildings can compound at scale. For example, in our projects, which were linked together in a district thermal energy loop, the results of an hourly analysis allowed one team of students to estimate that their design could provide enough chilled water to keep 280 houses in the nearby neighborhood cool at the peak of summer. Their design for a mixed-use housing building won the 2022 Commercial Competition.

Required Readings

I require students to read the entirety of the Design Challenge Competition rules and all standards referenced directly within the main text of the document.

Recommended Readings

Kwok, Alison and Walter Grondzik, *The Green Studio Handbook*, 2nd edition (Routledge, 2011).

Passive Building Design Guides for Multifamily (https://commercial.phius.org/sites/default/files/phius-passive-building-design-guide-multifamily.pdf) and Commercial Construction (https://commercial.phius.org/sites/default/files/phius-commercial-construction-design-guide.pdf)

Selections from Mang, Pamela, Ben Haggard and Regenesis Group, *Regenerative Development and Design: A Framework for Evolving Sustainability* (John Wiley, 2016).

Outreach Opportunities

I ask student teams to identify at least one community or industry partner. Students bear primary responsibility for initiating and maintaining the connection with the partner. Past partnerships have included community organizations focused on finance, urban livability, food access and affordable homeownership; mechanical engineering consultants and equipment manufacturers; photovoltaic panel manufacturers; prefabricated wall and floor panel providers; window manufacturers; and other high-performance architects, designers and builders.

Course Assessment

In 2018, we had one finalist team; in 2019, four; in 2020, we had three, and we won a first place award for our Mixed-Use Multifamily Building and an honorable mention for a retrofit in the Elementary School division. In 2021, both our two finalist teams won second place awards, one in the Urban Single Family housing division for a low-embodied carbon ADU design, and the other for a mass timber multifamily building. In 2022, we advanced three finalist teams, one of which won both the first-place prize in the Multifamily Building division and the Commercial Grand Winner prize. In 2023 and 2024, my colleague David Brubaker was the faculty lead with sole responsibility for a total of five finalist teams, including the 2024 Grand Winner award.

Looking Ahead

One repeated critique of the Solar Decathlon is a variation on this question: "Sure, it's great that we're teaching students to make more sustainable buildings, but do we really have to teach them to design *ugly boxes*?" A parallel, and equally stinging, critique is the suggestion that those of us who teach the Solar Decathlon are inculcating our students to become hapless tools of capitalism. Kiel Moe, writing about how the broader agenda of sustainability has been

FIGURE 14.3 A student from the 2022 Commercial Grand Prize Winner team with a model showing a portion of their proposed design.

defined, puts it this way: "energy can and should inflect the design of building, but not through technocratic training, a Calvinist ethos of minimization, or an ecologically dubious claim on net-zero anything." Moe calls for a new approach to "reconsider matter, energy and form with greater architectural ambition than what these closed concerns could ever afford alone."[9] The ugly box critique implies that architecture is opposed to mere building, whereas the embrocation of capitalism implies that *real* architecture operates without regard to economic concerns. These critiques show how hotly policed is the boundary around the field of architecture.

In closing, it is worth examining why the Solar Decathlon Design Challenge is likely to raise the hackles of architects like Moe. The competition is run by people and organizations (the Department of Energy and the National Renewable Energy Laboratory) that can be accurately and fairly described, at least to date, as technocratic. The Design Challenge was set up as a decathlon, so it consists of ten separate contests that are evaluated separately. "Architecture" is merely one of the ten, rather than the container holding all the categories together. The 2024 rules document defines the intent: "creativity in matching form with function, overall integration of systems and ability to deliver outstanding aesthetics and functionality both inside and outside the structure." Scoring criteria do not provide points for design or aesthetics on their own accord. This stands in contrast to the AIA's Framework for Design Excellence Criteria, which were developed by architects for architects. The Design for Integration criteria calls out a focus on "beauty and delight." In comparison, the Solar Decathlon rules document implicitly defines good architecture in a technocratic way: as "consideration" – fitting in, rather than standing out; "integration" – though in contrast to the AIA document, this means meeting, rather than surpassing expectations, and "quality" – paying attention to practical considerations, rather than privileging raw aesthetic experience. Evaluating architecture as a category with equal weight to life-cycle, health, community, or (shudder!) engineering constitutes a threat to both the borders around the field of architecture and its autonomy. These clashes with *doxa* can be difficult to see from outside the field. Given that a strength of the competition has been its appeal to teams from architecture, engineering and other disciplines, a challenge for the competition organizers is how to increase the legitimacy of the competition within the field of architecture while ensuring it remains accessible to those outside it.

More broadly, the fraught relationship of the Solar Decathlon and industry-facing sustainability programs with architecture curricula highlights an important choice for architecture education. Looking ahead, more investigation is needed of how the boundaries around the field of architecture might be explored, expanded and traversed. Sustainability competitions need not box in our buildings, our curricula, or our students.

Notes

1 Cécile Deer, "Doxa," in *Pierre Bourdieu: Key Concepts*, ed. Michael Grenfell (Routledge, 2014).
2 Michelle Laboy and Annalisa Onnis-Hayden, "Bridging the Gap between Architecture and Engineering: A Transdisciplinary Model for a Resilient Built Environment," *Building Technology Educators*' Society 1 (2019), https://doi.org/10.7275/MR5P-8X02.
3 Garry Stevens, *The Favored Circle: The Social Foundations of Architectural Distinction* (MIT Press, 1998).
4 Kathryn H. Anthony, *Design Juries on Trial: The Renaissance of the Design Studio* (Van Nostrand Reinhold, 1991).
5 Jonathan Bean, "Cracking the Code: A New Perspective on Architectural Education" (Association of Collegiate Schools of Architecture 110th Annual Meeting: EMPOWER, Los Angeles, 2022).
6 Mark C. Carnes, *Minds on Fire: How Role-Immersion Games Transform College* (Harvard University Press, 2018).
7 Karl Maton, *Knowledge and Knowers: Towards a Realist Sociology of Education* (Routledge, 2014).
8 Paul Hawken, ed., *Drawdown: The Most Comprehensive Plan Ever Proposed to Reverse Global Warming* (Penguin, 2017).
9 Kiel Moe, "Energy and Form in the Aftermath of Sustainability," *Journal of Architectural Education* 71, no. 1 (January 2, 2017): 88–93, https://doi.org/10.1080/10464883.2017.1260923.

PART IV
Design Build

15

BUILDING ON RESEARCH

A Hands-On Approach to Architectural Education

Joseph Wheeler

Course Overview

Course Name

SS:FutureHAUS AI; Industrialized Smart Housing

University / Location

Virginia Tech / Blacksburg, VA, USA

Targeted Student Level

- Undergraduate
  - All levels, all disciplines
- Graduate
  - All levels, all disciplines

Course Learning Objectives

In this course, students, across multiple disciplines, will:

- Survey areas of sustainable design and utilize the best means and methods for comprehensive, innovative design, fabrication and operation of a fully functional smart home.
- Work in teams with students and faculty across disciplines, as they ideate, design, fabricate and operate a cutting-edge prototype home that will be tested on a world stage.

DOI: 10.4324/9781032692562-20

- Learn from industry partners about products, materials and technology as a prototype is developed, constructed and tested for optimal performance and sustainability.
- Challenge the industry to adopt more innovative technologies and fabrication processes.
- Understand all aspects of sustainable design including carbon neutrality, green energy and efficient, industrialized, construction methods.

Course Learning Outcomes

Successful completion of this program will best prepare students for the "real world" by exposing them to the physical construction and performance of a competition prototype. Outcomes include:

- Students will learn to work across disciplines on complex problems.
- Students will be more prepared for the profession by designing and constructing a cutting-edge prototype that pushes the limits of current day technology.
- Students will be more prepared for the profession by having early collaborations with industry partners where they will be educated on current technology and have access to expertise of trade partners and technology manufacturers.
- By physically building the prototype, students will "learn by doing" enhancing their education through a hands-on approach with failures and successes fully visible.

Course Methodology

Philosophy

Maria Montessori, psychiatrist, educator and renowned pioneer in early educational philosophy, developed an effective method of teaching that builds on the way children learn naturally. A hands-on approach to learning through sets of tools that cover basic principles. She believed students, with the right material and direction, will be self-taught, driven by their own curiosity and interest. There are parallels to this method in vocational education that are also effective and widely in use. My courses embrace this philosophy as the students are engaged in realizing their own designs and learning through the hands-on approach to education, they also see the implementation of their work on a project that is physically realized and validated.

Traditional pedagogical tracks, especially in architecture, prepare the students for the profession without much exposure to real construction. In this course however, by getting students engaged in project-based learning, basic principles of design and construction are laid out in front of their own eyes. Lines on paper are translated to physical materials and details need to be worked out and made functional. At a higher level, much like work in the profession, students have access to consultants in academia and industry, and these become an additional branch of educational input.

FIGURE 15.1 The 2010 LumenHAUS, winner of the 2010 Solar Decathlon Europe Competition in Madrid Spain, utilized real time data from a weather station to drive the concept of responsive architecture, an early smart home driven by AI.
Source: Photograph by Jim Stroup

The young apprentices of the renaissance architects learned about practice by assisting their master architects with the design and construction of the great cathedrals and civic buildings. These apprentices learned from their mentors through the very process of building.

In a similar way, students in an academic setting, can work with their experienced professors and industry partners and learn through the physical construction of a high-level project.

Definition

Carbon neutrality – A goal that is difficult to calculate. It is more a common-sense approach to design where environmental sustainability, upfront carbon, passive design and renewable energy are considered when composing work of architecture. Decisions that can reduce carbon emissions and lower energy consumption should be automatically expected from any responsible designer.

Implementation

The Solar Decathlon is an international competition that challenges universities across the globe to design, build and operate sustainable, energy positive solar homes. The decathlon consists of ten events related to architectural design, engineering, home performance and sustainability. The competition has been held over a dozen times over the past 22 years demonstrating a wide range of innovative concepts for sustainable housing. Virginia Tech has competed in five of them, two of our most recent entries took first place in their respective competitions, one in Madrid, Spain and the other in Dubai, UAE. To win these competitions, teams with a wide range of expertise had to be formed and curriculum had to be adjusted to accommodate the highest levels of interdisciplinary education.

In 2003, after competing, and struggling, in Virginia Tech's first entry into the inaugural Decathlon event, a three-part pedagogical approach was developed, and continues to be a guide for further development with each successive Decathlon entry. The new multi-semester, multi-course, program was designed to take on a much more sophisticated type of project that could deliver high-performance homes that could compete with top universities globally.

FIGURE 15.2 The interdisciplinary program encourages teamwork between multiple disciplines. Pictured are students from Architecture, Mechanical, Computer and Electrical Engineering in The FutureHAUS "Dry Mechanical Closet."
Source: Photograph by Erik Thorsen Photography

Course 1: Research – Special Studies

The kickoff to our two-year program, and a critical course to its success, is a three-credit hour "special studies" research course open to all students, all years and all disciplines. In addition to providing valuable information through distinguished visiting lecturers, the students are directed to branch out and research a wide range of topics related to the sustainable house design including, but not limited to, renewable energy, sustainable materials, carbon neutrality, prefabrication, modular design, water conservation, aging in place, tele-medicine, accessibility and smart home technology. The final delivery from the course is a comprehensive research manual complete with options for technologies and materials available for the design team as the project is developed. The course usually attracts from 75 to 100 students across several disciplines and becomes the primary team-builder for the multi-year effort.

Course 2: Vertical Design Studio + Engineering Capstone

Through a vertical, graduate + undergraduate architecture design studio, the design process begins. The studio work is divided between one overall group project and individual studio projects, both ties to the solar decathlon design.

It is important to note that in all architecture design studios, it is critical to the student's design education for the individual to be in complete control of their own designs. This way, the individual student is directly responsible for their own design decisions and project development. This avoids the situation where one student's studio experience is not controlled by one or two of the strongest students in the class.

Parallel to the architecture studio, the undergraduate engineers, in collaboration with the architecture studio, are taking a cross-disciplinary senior level course to design/develop the engineering systems for the home. To accommodate cross disciplinary education, the College of Engineering has developed a course through the Engineering Education Department that allows students from Computer Science, Electrical, Mechanical and Industrial Systems Engineering to work together on a single capstone project.

Course 3: Vertical Design Build Lab + Engineering Capstone

The Spring semester is the "build" portion of the above listed vertical design studio and engineering capstone where a full-scale prototype with full integration of engineering systems is constructed and tested, this again, for the architects is in parallel to their own individual studio design project.

During the 1½ years leading up to this stage, the gift-in-kind sponsorships have been secured and the industry partners are ready to work with the students on installation of their products. By the end of this semester, the project will be ready for testing. Most of the core student team will remain on campus over the summer for paid internships to complete any unfinished construction and to fine tune all systems to ready the house for competition the next semester.

FIGURE 15.3 In collaboration with Siemens and Lutron, the LumenHAUS interdisciplinary team pioneered smart home strategies to optimize passive and active building performance strategies.
Source: Photograph by Jim Stroup

Performance Assessment and Analytics

In the design stages, the student's work is always verified by a wider range of analytical software including energy, structural, mechanical and acoustical simulation. In their green engineering course, life cycle analysis is performed as material selections are made for the built prototype. The best assessment however, in our case, is through the rigorous testing done through a week's worth of competition in the Solar Decathlon event. Buildings are monitored for energy production and consumption throughout the week of competition while performing all the major energy consumptive tasks in a residence, such as air conditioning and heating, cooking, dishwashing, showering and washing and drying clothes. Water consumption is monitored as well to encourage water conservation and recycling. The competition includes an event for teams to drive an electric car as many miles as possible utilizing their excess solar energy after the major house tasks are complete. Most miles driven on each team's vehicle wins that competition. The additional advantage to having both computer simulation and real-world testing, is that the latter verifies the accuracy of the computer simulation, furthering the student's education and confidence in their work.

Precedent Study

The popular phrase "*don't reinvent the wheel*" cannot be emphasized enough with Solar Decathlon projects. There are hundreds of prototypes from past competitions that have been constructed, tested and documented, that can serve as invaluable resources for the design team. In the first special studies course, students will collect invaluable information available about these houses and identify the strategies and technologies that have been proven successful, particularly the winners of overall events and specific competition categories. Professional precedents also contribute greatly to the research content. In the kickoff research course, students will survey these precedents and relevant technologies that will be worth considering for the new prototype. Topics of research include but are not limited to renewable energy technology, passive and active strategies, sustainable materials, building performance and energy efficiency, energy efficient appliances, prefabrication and industrialized manufacturing of buildings and products, water conservation. Notable discoveries are summarized and consolidated into a manual for the team's usage in the design stages later that year.

Project Support

Where university architecture programs don't typically have the financial resources to offset the costs of scholarships, prototype construction and transportation, or competition travel costs, we do rely on the university's assistance with facility support such as lab space, shop and fabrication equipment and in some cases even facility's trade expertise. We also rely on the university building inspector to verify the prototypes are professional and constructed per code.

Financially, the projects are supported through grants offered by the solar decathlon organization, by the architecture and engineering profession, and, most significantly, by industry sponsors. The industry supports by not only donating cash, but also with gift-in-kind material and product support, a win/win situation for both parties, since the products are showcased in the physical houses that are visited by tens of thousands of spectators as well as the communications outlets of each decathlon team.

University Facilities

In 1872, Virginia Tech was established as a land grant university, a federal program that contributed land to universities to promote agriculture, science and engineering. Since its establishment, Virginia Tech continues to support this practical teaching by providing facility resources for hands-on learning. Both the architecture and the engineering programs provide multiple research spaces with supporting wood and metal shops to facilitate the prototyping of research projects. The College of Architecture's Research and Development Facility (RDF) is the home base for the Solar Decathlon projects with high bay spaces for full scale building prototypes and supporting shops including a robotics and CNC laboratory that includes multi-access machining for research on industrialized methods for construction. Four full scale houses have been fabricated in this lab with support from secondary labs. The Mechanical Engineering's team utilizes their WARE Lab – for prototyping many of the mechanical systems, and the Electrical Engineering team utilizes both the Center for Power Electronics (CPES) Lab and the Future Energy Electronics Center (FEEC) lab.

Professional Partnerships

Often, the knowledge offered by the profession is just as important as financial support for the project. The design team relies heavily on the experience and intellectual resources of the industry professional. The introductory research course taps into that resource by inviting guest speakers to every class. These speakers share real-world expertise and/or demonstrate a wide range of technologies and materials that can be incorporated into the house designs. Those groups often continue to be engaged through the development of the project.

Course Evolution

After two major international wins, our pedagogical approach to the solar decathlon projects has evolved into a successful model, not only as a practical way to teach students through the design build process, but also as a way to produce competitive competition entries. What once was a long, drawn-out process of inefficient-development and much learning by doing, and redoing, now has evolved into an efficient three semester/two summer process that is directly informed by the industry. The introductory research course sets the tone, establishes the core interdisciplinary team and identifies the technology that will be utilized by the design team. The first semester vertical studio is used to design and perfect the design with direct collaboration of engineers, the second semester is for realizing the prototype and the final summer is for testing and fine tuning for competition entry. The SS: ARCH 4984/5984 course is the kingpin of the process, not only by identifying the students who will make the cross-disciplinary design team, but also to establish the direction of the research.

Below are two projects that were direct results of the three-course process.

The LumenHAUS 2010

The Virginia Tech LumenHAUS is a 680 sf, net-zero energy, solar powered house designed and built for the 2010 Solar Decathlon Europe Competition. In this concept, conservation is woven within innovation where natural ventilation, passive heating and daylight work in conjunction with new technologies that are sustainable and beautiful. The house, designed under the concept of "responsive architecture," where building components such as sun-shading devices, insulation panels and photovoltaic arrays can adjust to changing weather conditions based on information taken from a roof mounted weather station for optimal building energy performance. The small footprint of the pavilion-type house gives a functional generosity through the transformation of exterior walls and the integration of landscape and outdoor spaces. The house elements contract in extreme conditions and open to the expanse of sun and air in favorable weather. We can live in much smaller spaces while improving the quality of our dwellings. The environmentally sustainable house employs all tenets of green design and construction including the recycling of its own greywater and rainwater through a water-land landscape design, and sustainable materials that are recyclable, recycled, or are of low carbon and embodied energy. A geothermal heat pump system provided the most energy efficient way to heat and cool a home when the weather is too extreme for the passive strategies to be effective. The design is part of a larger concept called lumenocity, where, as an industrialized product, the home can grow as the family grows by adding modules. These components can be removed and sold if there is a need to downsize.

FIGURE 15.4 The passive strategy of the LumenHAUS allows the home interiors to expand out to its north and south decks during good weather.
Source: Photograph by Jim Stroup

Features of this house include:

- **A house larger than itself** – The pavilion plan, landscape integration and interior details allow a small house to function with elegance and efficiency; moveable components in the kitchen, office and bedroom transform spaces to multiple uses.
- **Eclipsis System** – The outermost fenestration is composed of stainless steel, laser cut sliding shutter shades providing sun protection, security, cross-ventilation, privacy, controlled views and a beautiful sunlit space. The second layer is a sandwich of aerogel filled polycarbonate insulation panels giving an R value of 24 and beautiful translucent daylight.
- **Alternative spaces** – In conjunction with the Eclipsis System, large sliding glass panels and curtain system offer differing spatial qualities; depending on the combination of active layers, the space can be altered between hard/soft, open/closed and intimate/public, with varying qualities of natural and electric light (including colors).
- **Landscape** – Fully integrated and seamless component of a comprehensive architectural expression; landscape serves for rainwater harvesting and grey water filtration.
- **Smart home** – The home weather station collects information to control shutter screens and insulation panels optimizing energy use and maximizing comfort.
- **iPad and iPhone interface** allows for personal control and feedback of energy use.
- **Geothermal heating and cooling** saves energy and is supported by passive heating (concrete floor/large glass exposure) and natural ventilation (Eclipsis System).
- **Radiant floor heat** – The best-quality heat; no moving air, no sound and greater efficiency.
- **Lighting of the future** – Dimmable Fluorescent and LED lights with digital control and sensors (daylight and vacancy sensors).
- **Putting light where it has not been** – Colored LEDs in insulation panels provide walls of multiple colors, no paint required; nighttime identity and security is established.
- **Enduring sustainability** – Materials chosen for their lasting sustainability and expressive capacity; IPE from certified renewable forests needs little maintenance, lasts forever and provides a beautiful floor reinforcing the sense of outside space; zero VOC paints; recyclable steel and aluminum.
- **Innovative transportation system** – An innovative transportation strategy allows the house to be easily transportable by incorporating front and rear hitches into the house frame that link with a rear dual axle wheel assembly and a front gooseneck for semi-truck attachment. This allows the house to be transported close to the road for overhead clearances. The transportation components are detachable and re-useable for economical transport of mass produced units.
- **Scalability** – The single-family small house can expand to larger plan types accommodating mass production for multiple units in a concept called LumenoCITY.

The team took first place honors at the international 2010 Solar Decathlon Europe Competition in Madrid, Spain. The house has been on public display on the National Mall and at the National Building Museum in Washington DC, in Times Square in New York City, on Millennium Park in Chicago, at the Farnsworth House in Plano, Illinois. It has welcomed more than 1 million visitors since its construction. It has also been featured on several major television networks and in international media outlets and received the National Institute Honor Award for Architecture by the AIA.

The FutureHAUS 2018

The Virginia Tech FutureHAUS research embraces the use of industrialized methods for fabricating buildings to accommodate current and future demands for more integrated and complex technologies in our homes. With pre-manufactured "cartridge" components, more complex systems can be delivered to our buildings to accommodate current and future market demands including user digital conveniences, energy efficiency measures, aging in place capability, accessibility for all age groups and continuous innovative modifications. This process also improves quality control in production and is in harmony with today's digital fabrication production methods.

FIGURE 15.5 The FutureHAUS won the 2018 Solar Decathlon Middle East in Dubai. The team pushed the concept of industrialized architecture to make a smart, prefabricated, solar home.

Source: Photograph by Erik Thorsen Photography

Designed and built by Virginia Tech students in 2018 to compete in the Solar Decathlon Middle East Competition, this wood frame home is prefabricated in twelve individual, compact, components or "cartridges" and is therefore easily transportable and installable with tight site restrictions. The team focused on energy efficiency, sustainability and water conservation. There was also a focus on smart interfaces, power management and energy monitoring. Though the structure and enclosure are wood construction (SIPS panels), much of the interior finishes are glass and prefinished modular MDF paneling offering alternatives to drywall on future modern homes. Exterior cladding is a white back-painted tempered glass rainscreen. Interior floors are cork backed back-painted white tempered glass. Ceilings are laminated patterned glass backlit by tunable led lights.

Features of this house include:

- **Flex space concept** – The home includes a home office wall, a bedroom wall and an automated murphy bed within the main living spaces that adjust to accommodate different size room needs at different times of day. Reducing overall square footage results in lower energy use and potentially lower house costs.
- **Aging in place and accessibility** – Various technologies incorporate easy to use interfaces that make living easier for the homeowners. The kitchen

FIGURE 15.6 Working with students from Interior and Industrial Design, The FutureHAUS team used "flex space" strategies to optimize space in a small house footprint.

Source: Photograph by Erik Thorsen Photography

incorporated a digital touchscreen backsplash and an island with an integrated smart monitor. Upper and lower kitchen cabinets can adjust to the user's needs. The bathrooms also have adjustable vanities and toilets. The 3D-printed sink base integrates three motion-controlled ports for cold, warm and hot water supply. Motion sensors throughout allow for slip-fall detection. All interior and exterior doors are automated for ease of use.

- **Building integrated photovoltaic glass (BIPV)** display featuring an automated drone hatch for future drone package deliveries.
- **Water conservation** – The shower incorporates an "orbital" filtration/recycling system that recycles shower water. The kitchen sink, bathroom sinks and washing machine recycle grey water through a landscape-based filtration system.
- **Glass fenestration** incorporates electronic tinting for building performance and BIPV for energy production.
- **Energy optimization software** monitors energy production and energy consumption paired with a home battery bank for load shifting guaranteeing optimal performance. Excess energy feeds back to the grid after recharging the home's electric vehicle.

After winning the international competition, along with many of the individual sub-contests, the house returned to the states where it was exhibited in multiple locations including Times Square in New York, the Virginia Tech "Innovation Campus" in Alexandria, VA, and the Drill Field on the Virginia Tech campus. In 2021, it was shipped back to Dubai to be featured as a "House of the Future" in the 2020 World EXPO. Sixteen students exhibited the house for six months where they welcomed over 1 million visitors from around the world.

Recommended Readings

Ching, Francis. *Building Construction Illustrated.* 6th ed. John Wiley, 2020.
Ford, Edward R. *The Details of Modern Architecture.* MIT Press, 2003.
Mazria, Ed. *The Passive Solar Energy Book.* Rodale, 1979.
Timberlake, James and Stephen Kieran. *Refabricating Architecture.* McGraw-Hill, 2003.
Zumthor, Peter. *Thinking Architecture.* Birkhäuser, 1998. https://archive.org/details/peter-zumthor-thinking-architecture/page/n31/mode/2up

Outreach Opportunities

The homes result in great outreach for the university. Many are published in books and magazines; however, the greatest exposure has been with the multiple post-competition exhibitions referenced above. All in all, the houses have been visited by well over 2 million guests interested in sustainability and renewable energy. The project continues to serve as a vehicle for university outreach and continues to spread the message that solar power is beautiful, affordable and responsible.

Course Assessment

In architecture, the vertical studio portions of the program cover the "integrative design" requirements for NAAB accreditation for both architecture graduate and undergraduate levels.

The syllabus for the course calls for the architecture students to:

- Select, analyze and evaluate a set of design precedents.
- Formulate effective responses to programmatic and situational conditions in the context of an architectural project using a variety of descriptive means such as constructive drawing, writing, modeling and analytic means.
- Employ and represent various constructive systems in the context of an architectural project in support of architectural ideas, habitation and place-making.
- Document an architectural project through the use of drawings, models, writings and oral presentations.
- Create work that analyzes, records and responds to existing contextual, regulatory, and environmental frameworks and scenarios.

FIGURE 15.7 After public exhibitions in Dubai, New York and Chicago, The FutureHAUS was shipped back to Dubai to be featured as a "House of the Future" in the 2020 World EXPO where the team welcomed over 1 million international visitors.
Photograph by Erik Thorsen Photography

- Participate in a critical dialogue about architectural works and the profession, substantiated by precedents and discipline-specific references.
- Evaluate and execute individual developments in areas of continuous interest, research and modes of study in architectural work.

In engineering, through each individual's cross-disciplinary capstone courses, the students satisfy their required senior accreditation obligations as well.

Looking Ahead

The 2025 Gateway Decathlon

Starting in the Spring of 2024, Virginia Tech will engage in our next Solar Decathlon competition, the 2025 Gateway Decathlon, a new competition hosted by a land developer in central St. Louis hoping to bring sustainability awareness to the future developments in St Louis. Seated at the base of the St Louis Arch, Virginia Tech's FutureHAUS AI will demonstrate the future of building utilizing AI technology in three primary areas: building performance, building/site design and home health. Following the model of automobile technology, the smart home will be automatically controlled to assist home users in positive ways that the students will determine as they develop the project. In a time where housing attainability is a huge concern, an exploration of prefabricated, factory-built housing is timely. The students will explore ways to use AI as a tool to assist in designing and configuring efficient floor plans and house plans to increase density without compromising quality.

16

LEVERAGING THE SOLAR DECATHLON COMPETITION AS A FRAMEWORK FOR A COMPREHENSIVE, COMMUNITY ENGAGED, CARBON NEUTRAL ARCHITECTURE STUDIO

Tom Collins

Course Overview

Course Name

Integrated Architecture Design Studio

University / Location

Ball State University / Muncie, Indiana, USA

Course Locations

- CAP: Indy Center, Indianapolis, Indiana
- Ball State University campus, Muncie, Indiana

Targeted Student Level

- Undergraduate
    - 4th year, Bachelor of Architecture, 5-year program
- Graduate
    - 1st year, Master of Architecture, 2-year program
    - 2nd year, Master of Architecture, 3-year program

DOI: 10.4324/9781032692562-21

Course Learning Objectives

In this course, students will:

- Understand the role of housing design in addressing climate change.
- Estimate onsite renewable energy generation to address operational carbon emissions in housing design.
- Model embodied carbon emissions by life-cycle stage.
- Apply third-party criteria to the design of carbon neutral housing.

Course Learning Outcomes

Successful Completion of this course will enable students to:

- Articulate a cohesive approach to addressing the climate change and environmental impacts of housing design.
- Design housing that addresses operational carbon emissions using onsite renewables.
- Design housing that considers the embodied carbon emissions at various life-cycle stages.
- Design carbon neutral housing using third-party criteria.

Course Methodology

Philosophy

Building materials, construction and operations represent 39% of carbon emissions worldwide.[1] Through the design of buildings that reduce operational and embodied energy usage, students in schools of architecture are simultaneously reducing greenhouse gas (GHG) emissions associated with that energy. When buildings are designed to be net-zero energy, they produce at least as much as they use onsite with renewables. The exploration of new sustainability approaches to building design in studio relates directly to addressing global climate change.[2] When buildings are designed to be all-electric in regions like the Midwest, where electricity generation is carbon intensive,[3] grid-connected, net-zero energy buildings add renewable power generation and anticipate the region's transition to a clean energy economy. While net-zero buildings represent a significant paradigm shift and challenge to status quo building design in the profession and the academy,[4] they must go even further to holistically address the role of carbon neutral buildings and sustainability.[5] A body of literature argues in favor of mainstreaming carbon neutral design in studio curricula.[6]

Ball State University uses an integrated, community-engaged, immersive-learning approach coupled with the US Department of Energy's (DOE) Solar Decathlon (SD) competition and framework[7] to guide and inform carbon neutral design curricula in two upper-level integrated design studio courses. We

FIGURE 16.1 Completed Alley House in Indianapolis.
Source: Photograph by author

used the SD Build Challenge variant of the competition on our recent 2-year design/build project called the Alley House.

To address operational energy and carbon emissions, our teaching pedagogy relies on a modified three-tier approach to sustainable design long promoted by Norbert Lechner[8] where students first focus on cost-effective strategies to reduce envelope heat loss/gain and optimize for site orientation (Tier 1). Tier 2 applies bioclimatic passive design strategies to address heating, cooling, lighting and ventilation. Midwest buildings still require active systems to address remaining loads, but Tier 3 aims to dramatically reduce the size, cost, complexity and space requirements of active systems due to the energy efficiency measures employed in Tiers 1 and 2. We consider renewables to be an additional Tier 4 because they are applied once the building energy loads are reduced. Using the SD framework, students design for net-zero energy using performance simulation, performance metrics (such as EUI or HERS Score) and industry benchmarks.

To address embodied energy and carbon emissions, our teaching pedagogy relies on a life-cycle assessment (LCA) approach that considers and tracks the impacts of building materials production, construction, use, end of life and reuse with a focus on circularity. Student design teams are not expected to achieve net-zero carbon for the materials and building construction phases of the life cycle but are encouraged to find and use low-carbon materials and to consider emissions reductions associated with materials made locally, that minimize construction waste and that have a long lifespan. Biobased materials such as wood products are encouraged because they offset emissions through carbon sequestration and have lower embodied carbon during processing. Teams are encouraged to reduce the use of higher-carbon materials through reuse or materials with recycled content. Using the Solar Decathlon framework, students always design for embodied environmental impact using LCA tools, predicted performance metrics (such as GWP) and industry benchmarks.

The Solar Decathlon framework includes requirements that inform the student projects, but still allow teams flexibility and agency in the design process.[9] The competition and framework function as what Holzer et al. describe as a third party studio leader to help students design holistically.[10] The competition also aligns with the culture of hands-on, learning in our college. Students work in collaborative teams, which allows for diversity of ideas, peer teaching and more detailed output. The Build Challenge is a housing competition, and this scale is appropriate for integrated design. Projects involve community partners, which provides a real-world experience critical to student learning.[11] Partners bring additional requirements, goals and aspirations, which help our projects stand-out among the competition.

The community partner for the Alley House was a community development corporation focused on neighborhood revitalization, affordable housing, quality of place and educational opportunities. Student teams sought synergies between operational and embodied carbon and the organizational concerns of the partner such as the positive impact of reduced monthly utility bills on low-income residents to achieve cost, energy and carbon savings. An important aspect of carbon neutral design for our students is understanding and reconciling barriers associated with innovative solutions to energy and emissions reductions within status-quo design/construction practices in our region.

Definitions

Carbon neutrality – Operational energy use reductions that enable a building to meet remaining energy demands through onsite, renewable energy systems using no onsite fossil-fuel based energy.

Bioclimatic design – A "philosophy of design where the building operates in concert with the prevailing climate, its occupants and its program."[12]

Immersive learning – High-impact learning experiences that involve collaborative student-driven teams, guided by faculty mentors.

Net-zero – "One hundred percent of the building's energy needs on a net annual basis must be supplied by on-site renewable energy. No combustion is allowed."[13]

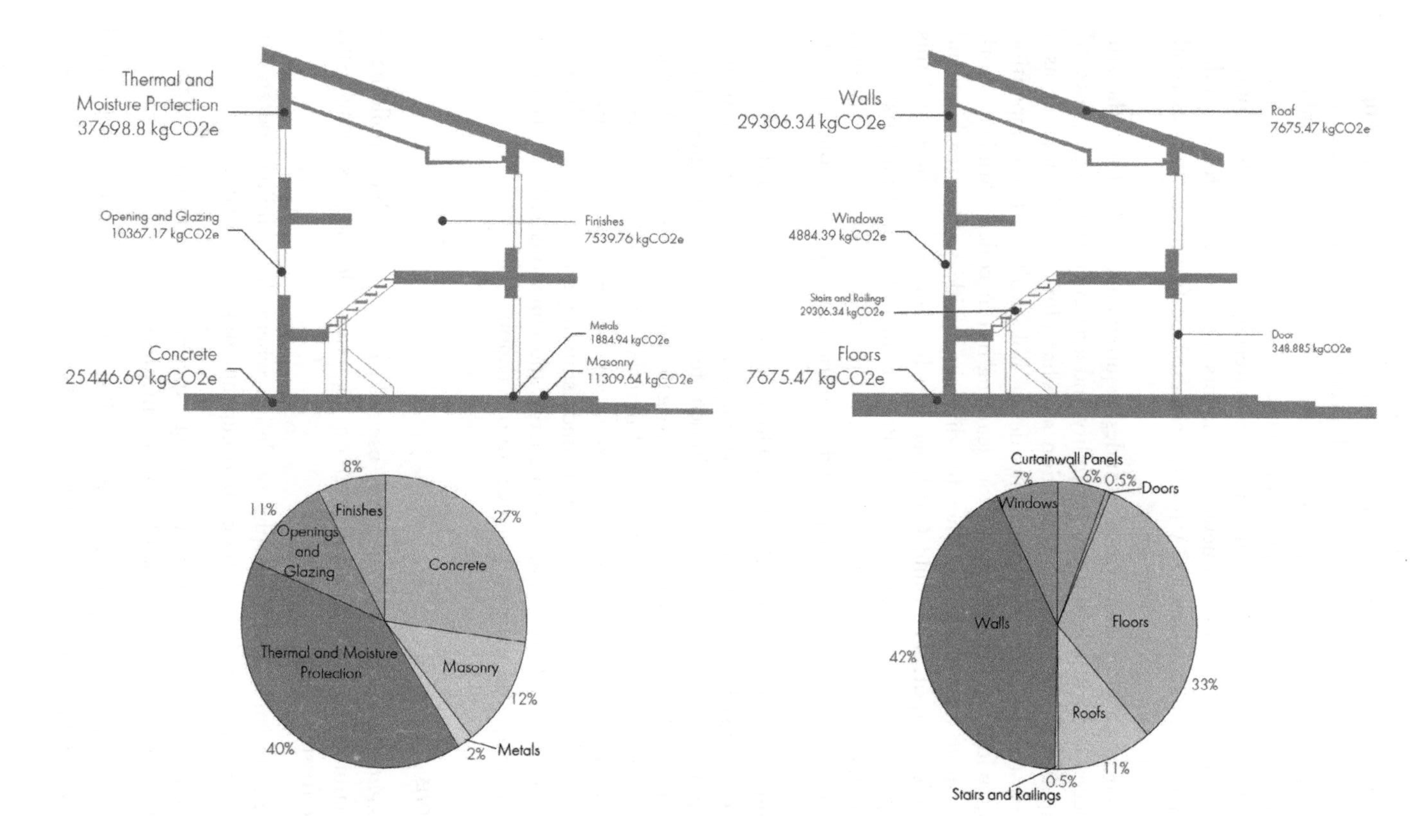

FIGURE 16.2 Embodied carbon study for the Alley House project.

Implementation

General

Precedent Studies

Precedents of carbon neutral housing are limited. Each semester, students study projects in several categories: past SD projects (Design and Build Challenges), net-zero energy housing, PHIUS certified projects and Zero Energy Ready Home projects. There is a gap in the literature related to affordable, carbon neutral housing, and our precedent studies ask students to examine precedents through an "affordability lens" even if it wasn't a stated goal.

Design Concept

The SD requirements provide a pedagogical framework for our studios, but all competing teams use this same framework. We build upon the requirements to address core ideas related to local issues including affordable housing, neighborhood development and innovation in the building industry. Strong ideas and local relevance drive successful SD projects, and this helps us distinguish our schemes from other teams that employ similar systems and strategies. The competition acts as a framework, but not a roadmap or a checklist. The framework is flexible enough to accommodate a variety of curricula.

Facilities

Having an academic facility near the design/build site allows students to make more site visits and for partners to visit the studio. When the team members are far from the site, trips are planned in advance and construction schedules may change. On the Alley House, the graduate students were in Indianapolis and the undergraduates were in Muncie 60 miles away. To address this challenge, we used off-site fabrication to build components in a shop that were installed at the site later.

Evolution of the Course Over Time

The Alley House was Ball State's first solo SD Build Challenge project. The college has engaged in the Design Challenge since 2017, but there are significant differences with a design/build. Thus, we only have one Build competition experience. However, we adjusted the studio courses each semester throughout the process to address challenges and are now discussing ways the courses can evolve for our next design/build.

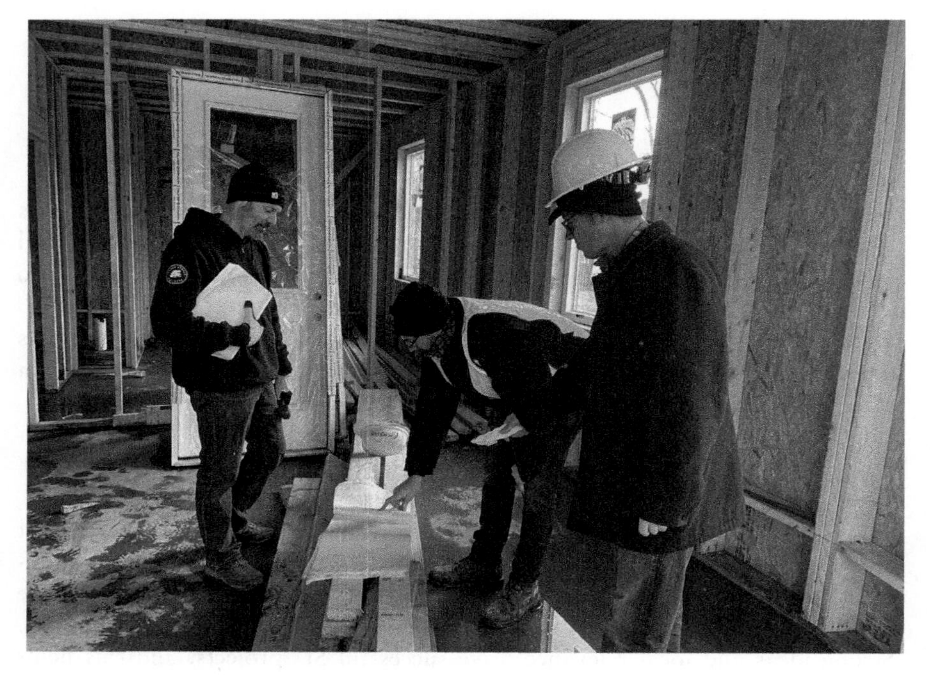

FIGURE 16.3 Students at the Alley House with the general contractor.
Source: Photograph by Pam Harwood

Assignments

The studio courses have three types of assignments: briefs, competition submissions and design reviews. Briefs are weekly assignments that introduce students to a variety of concepts and tools and are tied to course objectives and the SD framework. Students apply complex integrated design considerations to their scheme each week. Since the briefs focus on 1–2 topics, integration occurs through an incremental and cumulative process. The briefs introduce content in manageable bites. While each brief adds new design considerations, teams evaluate the impact on previous design decisions weekly. Briefs include an individual and a group component, which allows students to have a percentage of their course grade that is not shared with the group. Briefs include upcoming studio schedule updates, a rationale for the design considerations introduced, assigned tasks, links to resources and relevant NAAB criteria.

The SD Build competition has its own schedule that generally aligns with academic calendars and includes team submissions at set intervals throughout a two-year cycle. The briefs allow instructors to maintain consistency with how we deliver carbon neutral design experiences to students even when the competition submissions vary each semester. The submissions tie directly back to the SD framework and ten contest areas, which allows teams to document and reflect on the project holistically. Competition materials are assessed by subjective feedback and objective scores from jurors/organizers.

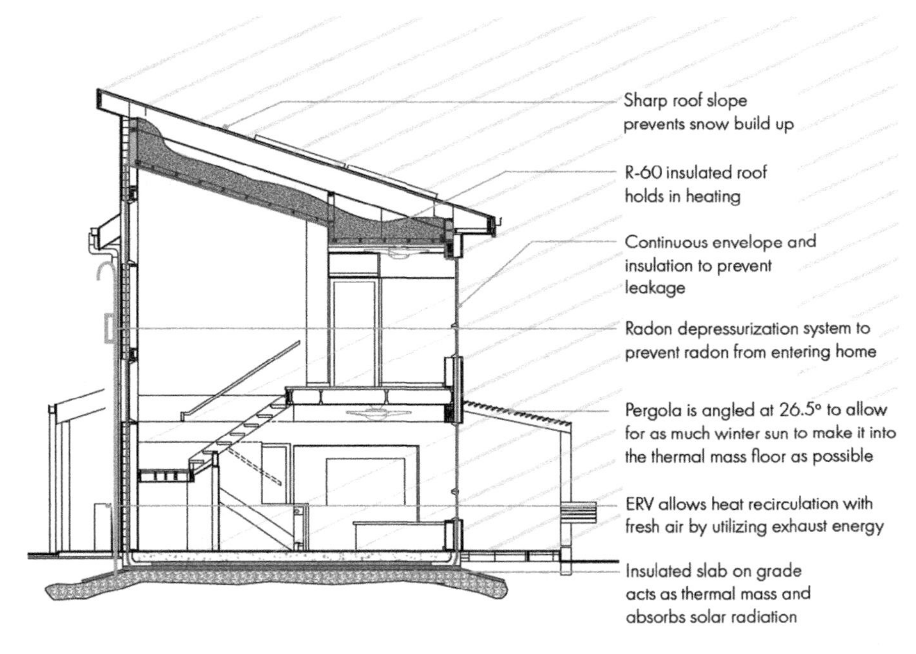

FIGURE 16.4 Integrated design diagram for the Alley House.

FIGURE 16.5 A student presenting a physical model of the Alley House.
Source: Photograph by Vern Slocum, NREL/Solar Decathlon

We schedule four design reviews each semester at one-month intervals, which are cumulative and previous materials are edited/revised for subsequent reviews. The materials become more detailed as the semester progresses. Review formats may be panel juried or "gallery-style." Each Spring Semester, the SD hosts a juried event at NREL in Golden, CO. A rubric is used to assess completeness of deliverables, presentation quality and engagement with integrated design.

Analytics

Analytical tools are used throughout the courses to inform design decision-making. For early to mid-stage building performance modeling, we use cove.tool software, which is available to students through an educational license. We find that there is a minimal learning curve for students with this software and that the interface is clear and easy to understand. Teams use cove.tool to track the progression of performance metrics like EUI, sDA and ASE.

For life-cycle assessment (LCA), we use Tally software, which has a slightly steeper learning curve, but allows our students to easily obtain useful GWP metrics. Tally is introduced around mid-term, which is when students are typically ready to model their schemes in Revit, which is required to use the Tally plug-in.

SD provides and recommends a suite of other analytic tools for teams including BEopt, RSMeans for cost analysis and REM/Rate for HERS analysis. They also provide access to Revit Cloud work-sharing, which is not included in the Revit educational license and is useful for team collaboration.

Several excellent free analysis tools are used for more discrete analysis work such as Climate Consultant, PV Watts Calculator, Opaque, HEED and Therm. Finally, we use the PHIUS Core Prescriptive web-based SnapShot tool to recommend climate-specific envelope values for our location and to comply with the standard for third party certification.

Four of the ten SD contest areas use measured data to determine points earned by teams. The DOE sends teams a kit of monitoring tools, assists with the set-up via webchat and collects the data remotely. Teams access the real-time data through a web portal to track the performance for various performance tests. The Alley House units have additional monitoring for ongoing energy, IAQ and solar production.

Required Readings

Corner, Donald, Jan Fillinger and Alison Kwok. *Passive House Details: Solutions for High-Performance Design*. Routledge, 2017.

King, Bruce. *The New Carbon Architecture: Building to Cool the Climate*. New Society Publishers, 2017.

Recommended Readings

Grondzik, Walter T. and Alison G. Kwok. *Mechanical and Electrical Equipment for Buildings*. Thirteenth ed. Hoboken, New Jersey: John Wiley & Sons, 2019.

Iano, Joseph and Edward Allen. *The Architect's Studio Companion: Rules of Thumb for Preliminary Design*. Seventh ed. Hoboken, New Jersey: John Wiley & Sons, Inc, 2022.
Kwok, Alison G. and Walter T. Grondzik. *The Green Studio Handbook: Environmental Strategies for Schematic Design*. Third ed., Vol. 1. New York, New York: Routledge, 2018.

Outreach Opportunities

Student Advocacy

Design/build projects in design studios provide formative experiences for students. Faculty advocate for students throughout the process by providing various kinds of support to help students achieve their best work including connecting teams with resources and experts; providing feedback through document edits and desk critiques; mentoring students on how to market their experience to potential employers; and helping to maintain morale.

Professional and Community Connections

Carbon neutral design/build projects are complex, and students rarely possess the necessary expertise to address the various design concerns such as HVAC design, financing, or regulatory compliance. Professional and community partnerships help close this knowledge gap and offer opportunities for students to form professional and community connections. However, developing partnerships takes time and professionals and community members are busy people. It is challenging to rely heavily on volunteer support for projects that are under time constraints. Hiring professionals can be the best option, but there are fewer learning opportunities for students in this scenario.

Conferences and Publications

An important part of our teaching pedagogy in the integrated design studios is documenting and disseminating the outcomes of the projects. We continue to seek venues outside the university to share our experiences, process and lessons learned. Recent conference presentations and papers were include PhiusCon 2023 in Houston, EDRA54 in Mexico City, and the 7th RBDCC at Penn State University.

Course Assessment

NAAB Accreditation

Our integrated design studios play a critical role in NAAB accreditations. Under the NAAB 2020 Conditions,[14] the studios contribute to program criteria such as PC.6 Leadership and Collaboration and satisfy student criteria such as SC.1 Health, Safety and Welfare in the Built Environment, SC.5 Design

FIGURE 16.6 Community groundbreaking for the Alley House.
Source: Photograph by author

Synthesis, and SC.6 Building Integration. Since SC.5 and SC.6 are the only criteria that require evidence of student work, there is added pressure placed on our integrated design studio projects. The 2020 Conditions do not explicitly refer to energy or carbon and allow programs to determine how to implement and assess the criteria. Thus, the accreditation criteria alone are insufficient to maintain consistency across schools of architecture related to carbon neutrality in design, which is concerning 14 years after the 2010 Imperative aimed at studio curricula.

Competition Achievements

Ball State's Alley House project won the 1st Place Overall award at the 2023 Solar Decathlon Build Challenge competition. The project also won various contest area awards including 2nd Place for Architecture, 2nd Place for Market Analysis, 2nd Place for Integrated Performance, 1st Place for Occupant Experience, 1st Place for Energy Performance and 2nd Place for Presentation.

Other Awards

The Alley House received a 2022 State Farm Insurance Neighborhood Assist Grant, a 2023 Society of American Registered Architects (SARA) National Design

FIGURE 16.7 Team reaction to the award announcement at the Solar Decathlon Event. Source: Photograph by Vern Slocum, NREL/Solar Decathlon

Award, a 2024 American Institute of Architects (AIA) Indianapolis Chapter Design Excellence Award and a Ball State University Immersive Learning Faculty Award. The Alley House also received PHIUS Zero 2021 certification.

External Examiners

Student teams participating in the SD Build Challenge competition interface with external examiners throughout the process. DOE staff review submissions, provide feedback, meet with students virtually and even make in-person site visits. At the annual SD competition event, students present to juries with expertise in a variety of areas including architecture, engineering, sustainability, codes, construction methods and building performance. Studio faculty also invite professionals and alumni to attend milestone design reviews.

Measurement of Student Outcomes

The SD competition juries provide teams with objective scores in each of the ten contest area categories. Studio faculty assess team progress and individual student achievement through graded weekly briefs, design review materials and competition submission reports. Therefore, measurement of student outcomes occurs internally within the studios and externally from the competition.

FIGURE 16.8 Ball State team presenting at the Solar Decathlon Event.
Source: Photograph by Vern Slocum, NREL/Solar Decathlon

Student Reactions

Undergraduate Navy Lynch commented that "the SD framework challenged me to explore how energy is generated, stored and used as well as take a deeper dive into conserving energy through sustainable design." Graduate student Emily Rheinheimer commented that SD "provided a structured framework to rapidly introduce students to thinking beyond basic building standards." Finally, graduate student Alejandra Lagunas commented that the SD "framework can benefit [a] student's education because it goes above the standard curriculum topics. Participation … was crucial to my education because I developed new skills and knowledge about net-zero and clean energy design techniques."

Lessons Learned

Design/build projects span academic semesters, and our experience suggests that providing some student continuity across courses is important. Also, sites far from the studio classroom limit hands-on construction time for students. Off-site fabrication of components does not replace onsite engagement. Design/build projects are intense experiences for students and frustration, burn-out and stress are issues faculty must manage to meet expectations, goals and timelines. Building student designs is expensive and considerable faculty effort and time go toward fundraising and securing sponsorships/donations. At the end, students feel immense pride and accomplishment, and many report that it was a seminal experience in their education.

FIGURE 16.9 The team onsite learning about control layer continuity.
Source: Photograph by Pam Harwood

Conclusions

Ball State University uses the Solar Decathlon design/build competition and framework to inform integrated architecture design studios focused on carbon neutral housing. Projects address local issues such as affordable housing, urban revitalization, bioclimatic design, antiquated construction practices, regulatory barriers and local materials. The objective is to use creative problem solving, design thinking and industry best-practices to inspire and catalyze the building industry in the Midwest toward lower carbon ways of building.

Looking Ahead

The SD competition provides a useful framework for integrated, carbon neutral design in architecture studio courses even if students do not participate in the competition. The Alley House will become a precedent for future teams. Continuous monitoring of the units also allows students to learn about predicted vs. actual performance. Occupant feedback from tenants will also be valuable to future teams. Since all new SD Build projects will likely be on local sites, this will allow Ball State to work with different communities each cycle to raise awareness of carbon neutral design practices. Student

FIGURE 16.10 Members of the team at the ribbon cutting ceremony.
Source: Photograph by Robbie Mehling/Ball State Office of Immersive Learning

teams are designing and building a house, but they are also providing a future-focused model of design in our Midwest region.

Notes

1 Amy Cortese, "The Embodied Carbon Conundrum: Solving for All Emission Sources from the Built Environment," New Buildings Institute, December 16, 2022, https://newbuildings.org/embodied-carbon-conundrum-solving-for-all-emission-sources-from-the-built-environment/.
2 Eray Bozkurt, "Exploration of Climate Change in Architectural Design Studio." In *SHS Web of Conferences* 48 (2018): 01039.
3 US Energy Information Administration (EIA), "Carbon Intensity of US Power Generation Continues to Fall but Varies Widely by State," November 11, 2023, www.eia.gov/todayinenergy/detail.php?id=53819
4 Mark Olweny, "Leadership Is Critical in Mainstreaming Sustainability in Professional Education," *Building and Cities* (2021); Robert Grover, Stephen Emmitt and Alex Copping, "Critical Learning for Sustainable Architecture: Opportunities for Design Studio Pedagogy," *Sustainable Cities and Society* 53 (2020): 101876.
5 Sergio Altomonte, Peter Rutherford and Robin Wilson, "Mapping the Way Forward: Education for Sustainability in Architecture and Urban Design," *Corporate Social Responsibility and Environmental Management* 21, no. 3 (2014): 147.
6 Fionn Stevenson and Alison Kwok, "Mainstreaming Zero Carbon: Lessons for Built-Environment Education and Training," *Buildings and Cities* 1, no. 1 (2020): 687–696.
7 US Department of Energy, *2023 Build Challenge Rules*, Washington, DC: US Department of Energy, 2023.

8 Norbert Lechner and Patricia Andrasik, *Heating, Cooling, Lighting: Sustainable Design Strategies Towards Net Zero Architecture*. Fifth ed. Hoboken, New Jersey: John Wiley & Sons, 2022; B. Haglund, "Pioneering the 2010 Imperative in Studio: Carbon-Neutral Studios 2006 and 2007," PLEA 2008 – 25th Conference on Passive and Low Energy Architecture, Dublin, October 22–24, 2008.
9 US Department of Energy, *2023 Build Challenge Rules*, Washington, DC: US Department of Energy, 2023.
10 Dominik Holzer, Lu Aye, Brendon McNiven and Lucy Marsland, "Lessons from Integrated Design Studios: Focusing on Zero Carbon," Australian Institute of Refrigeration Air Conditioning and Heating, i-Hub Summit, 2020.
11 Robert Grover, Stephen Emmitt and Alex Copping, "Critical Learning for Sustainable Architecture: Opportunities for Design Studio Pedagogy," *Sustainable Cities and Society* 53 (2020): 101876.
12 Sergio Altomonte, Peter Rutherford and Robin Wilson, "Mapping the Way Forward: Education for Sustainability in Architecture and Urban Design," *Corporate Social Responsibility and Environmental Management* 21, no. 3 (2014): 147.
13 International Living Future Institute, "Zero Energy Certification Overview," 2022, https://living-future.org/zero-energy/certification/
14 National Architectural Accrediting Board, "Conditions for Accreditation, 2020 Edition," 2020, www.naab.org/wp-content/uploads/2020-NAAB-Conditions-for-Accreditation.pdf

17

TIMBER TECTONICS IN THE DIGITAL AGE

Nancy Yen-wen Cheng

Course Overview

Course Name

Timber Tectonics in the Digital Age, http://timbertectonics.com

Universities / Locations

- University of Oregon / Eugene, Oregon
- Oregon State University, Corvallis, Oregon

Targeted Student Level

- Undergraduate
    - 4th and 5th year, Bachelor of Architecture, 5-year program (University of Oregon)
    - 4th year, Bachelor of Science, Wood Innovation for Sustainability, Architectural Engineering or Construction Engineering 4-year programs (Oregon State University)
- Graduate
    - 2nd year, Master of Architecture, 2-year program (University of Oregon)
    - 3rd year, Master of Architecture, 3-year program (University of Oregon)
    - 1st year, Master of Science, Wood Science, Civil Engineering or Construction Engineering 2-year programs (Oregon State University)
    - 1st year, Doctorate of Philosophy, Wood Science, Civil Engineering or Construction Engineering 3-year programs (Oregon State University)

DOI: 10.4324/9781032692562-22

Course Learning Objectives

In this course, students will:

- Study structural systems and their use in contemporary timber construction practice.
- Learn how to design and analyze parametric variations of timber structures.
- Practice integrated AEC collaboration workflows and visualize construction processes.
- Learn design implications of how wood can be engineered, manufactured and assembled.

Course Learning Outcomes

Successful completion of this course will enable students to:

- Design and analyze parametric variations of timber structural systems.
- Develop fabrication solutions to support architectural design.
- Collaborate in integrated design teams.
- Document design solutions, visualize construction processes and communicate to different audiences.

Course Methodology

Philosophy

This course provides students with technical and interpersonal skills needed to be effective in a professional setting. Architects and engineers need to collaboratively design with physical material properties and structural principles using current technologies. With growing demand for renewable materials, students need to understand engineered wood products, fasteners, pre-fabrication and on-site methods. The Timber Tectonics in the Digital Age studio addresses these industry needs via interdisciplinary digital collaboration focusing on wood design and construction. With live instruction connecting remote universities, it fosters understanding of communication and negotiation processes, immersing students into workflows for AEC collaborative design, simulation and fabrication. Within an 11-week quarter term, it focuses on structural design and material efficiency. The course coaches students in parametric design with structural simulation to complement more intuitive sketching and physical prototyping. This chapter explains the course objectives and evolution.

Definitions

Carbon neutrality through material efficiency – To achieve Carbon Neutral Design, the class focuses on renewable timber resources, structural efficiency

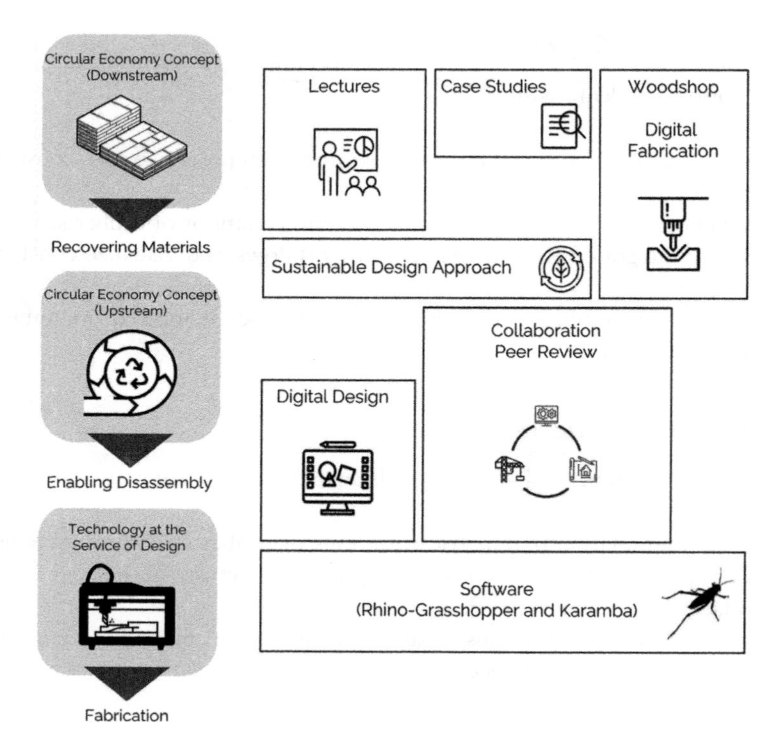

FIGURE 17.1 Underlying principles and course activities.[1]

and Design for Disassembly and Reuse (DfDR). Our class is situated in Oregon, where students are surrounded by wood frame buildings, a growing mass timber industry and marketing. Because tallying wood's carbon impact depends on assumptions about end-of-life disposal for biogenic carbon, we don't emphasize life-cycle analysis. Instead, we coach students on finding appropriate wood products for different tasks, optimizing form and dimensions to encourage reuse and reduce waste.

Digital collaboration – While architectural professionals commonly work with varied internal and external teams, students typically have few opportunities to bridge disciplinary boundaries for teamwork. This class asks them to engage, listen to and negotiate with two to four partners, spanning two Oregon locations: Eugene and Corvallis. To connect over fifty miles, the course uses Canvas Content Management System, Zoom video conferencing, Miro whiteboard, Microsoft Teams group text and file sharing. Activities promote crucial interaction skills, such as restating partner messages, providing positive feedback and negotiating conflict.

Interoperability – To participate in a digital team workflow, professionals need to share high-quality building information such as 3D models (in our case McNeel's Rhinoceros with Karamba structural simulation) and understand their partners' capabilities. For smooth data exchange or interoperability, they should be able to open and work on the project using a shared file format without losing information. To use a different software, they must be able to import the shared file OR request information in a format they can use, and then after working on it, communicate results that can be used by others.

Reciprocal frame – We can increase the spanning capability of short members using a mutually supporting network or circuit, known as Reciprocal Frames. This structural type originated in medieval times and has re-emerged in times of material shortage. Examples include the tepee, Leonardo bridge and the Zollinger lamella wall.[2]

Implementation

The course began in 2017 along with the Tallwood Design Institute partnership between the University of Oregon (UO)'s College of Design and the Oregon State University (OSU)'s Colleges of Forestry and Engineering. Architect and engineer Dr. Mariapaola Riggio of OSU's Wood Science and Engineering department provides timber design, construction and technology insight. I bring digital architectural design and collaboration expertise.

Assignments

Format and Content

The course introduces students to a structural system, shows how to parametrically model variations and then guides small teams in designing, detailing and prototyping a specific project with that system (Figure 17.2). It began as a broad overview of timber structural systems and has moved towards intensive engagement of a single system to support full-scale prototyping. To address changing needs, the brief asks students to design demountable temporary shelter. The changing program brief has addressed local needs such as gathering, performance and adaptive housing, with recent consideration of recycling.

We begin with a course introduction and team building. To initiate rapport at a face-to-face kick-off, each student interviews another to write a biographical profile, and pairs work together on an activity. After this kick-off session, students virtually share each other's profiles on a talent marketplace, from which students propose teams. Within a few days, they form teams, and fill out planning worksheets to support future team interactions. The Team Agreements foster awareness of different talents, schedules and conventions. Later, they jointly fill out a Project Management worksheet with concrete objectives and timetables.

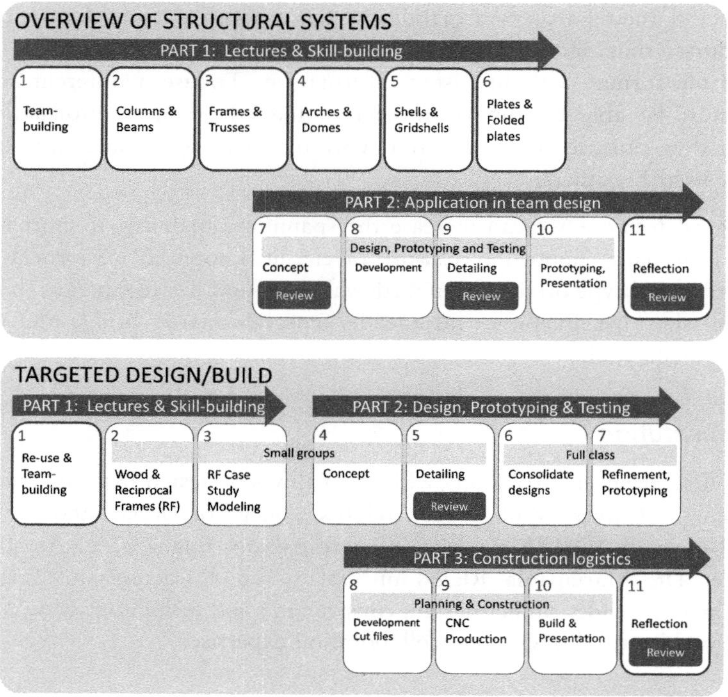

FIGURE 17.2 Course structure has evolved from an overview of many structures to a targeted design-build studio.

In the first third of the term, the students study structural forms, Grasshopper parametric design and Karamba analysis techniques to simulate the structural forms. They modify given digital examples or model case studies to see how varying dimensions, materials and loads affect stresses. As the term progresses, they apply lessons into the design project, creating physical and digital prototypes.

Online lectures explain the force-carrying mechanisms using basic forms and reveal growing sophistication towards contemporary examples. Video tutorials show how to digitally model adjustable structures with linear elements (beams, columns, frames, trusses, arches) and then 3D surface structures (vaults, domes, plates, folded plates, with form-finding for shells and gridshell).

Deliverables

Students are asked to regularly share inspirations, design work in progress with digital experiments and reflections on a digital class whiteboard. Periodically, they hand in more formal ePortfolios using a PowerPoint template file that asks for learning objectives, reflections on progress made, feedback for instructors and next steps. Major design presentations are summarized on a public blog and with group discussions about conflict and negotiation styles.

Over seven annual iterations, we have moved towards intensively engaging students in a narrower domain through collaborative design in a studio format. We originally pushed all students to have consistent baseline skills and knowledge in parametric design, structural optimization, timber construction. Now we accept that asymmetric skill specialization can create more effective teams, as seen in professional situations.

Within small groups, individuals take on design, analysis, representation and communication tasks according to abilities. Experienced engineering students can provide structural advice and follow with quantitative analyses. They request data in a format they can use, even if that means translating multiple static versions from a parametric model. More motivated and skilled students can take on outsized roles, so if teams do not have a natural manager, instructors need to ensure that all students have tasks that suit their interests and training.

The 2020–2021 global pandemic accelerated a move towards a "flipped" classroom with pre-recorded lectures and tutorials, emphasizing class time for live interaction. Small group conferences and individual reflective reports reveal ongoing efforts and competencies, In the large group, interactive polling using Zoom chat or Mentimeter increases accountability.

Material Awareness and Design for Construction

To understand wood design, students need to see and experience how the directionality of wood fiber impacts the behavior and performance of timber products at scales of the connection, room and building. Hands-on woodshop exercises can develop this understanding. Creating furniture joints can familiarize students with machining and woodworking, and adapting them to architectural scale can be challenging. The context makes joint studies more meaningful. Student motivation and learning are stronger when they work on joints for their own design project.

Modeling Construction Case Studies

Students who start with a timber structure have had more success than those who start with a digital vision and must adjust the form for stability. One team[3] was able to mimic the modular construction of the Trondheim gridshell[4] to build a more organic form (Figure 17.3). Its use of inner and outer layers of flexible wood laths connected by fasteners in slots allows the flat element matrix to be pushed and locked into a vaulted shape. Creating their own version of the modular system, they were able to construct a large rectangle and then reconfigure the same components to create a demonstration arch.

FIGURE 17.3 The modular Trondheim gridshell system supports easy reconfiguration.

Collaboratively building with small scale components can build trust and accelerate a deeper understanding of structural systems. Physical models efficiently communicate design intentions and address structural and constructability issues without digital challenges. In Fall 2023, when we required physical models of case studies, we wondered whether mimicking would stifle creativity. But scale modeling a well-documented component system provides a robust tactile tool kit for free exploration, and steers students away from problematic unstable virtual visions. Our students enjoyed learning through hands-on design and construction experiments, starting with 3 mm plywood (one-sixth of ¾″ plywood) and iteratively moving to full-scale.

Design speculation depends on modeling craft to yield understanding. For example, initial small-scale studies of reciprocal frames based on triangular tessellations seem to work fine when laser cut out of thick cardboard, but larger wood models showed that the perpendicular laser cuts are poor simulations of the diagonal notches required. The challenge of cutting complex diagonal notches led us in Fall 2023 to focus on only using cuts perpendicular to the plywood surface, as they can be accurately and efficiently prototyped with laser cutting and computer numeric controlled (CNC) cutting.[5]

In a structure composed of many mutually dependent elements, the stability of the overall structure depends on many joints. At the tabletop scale,

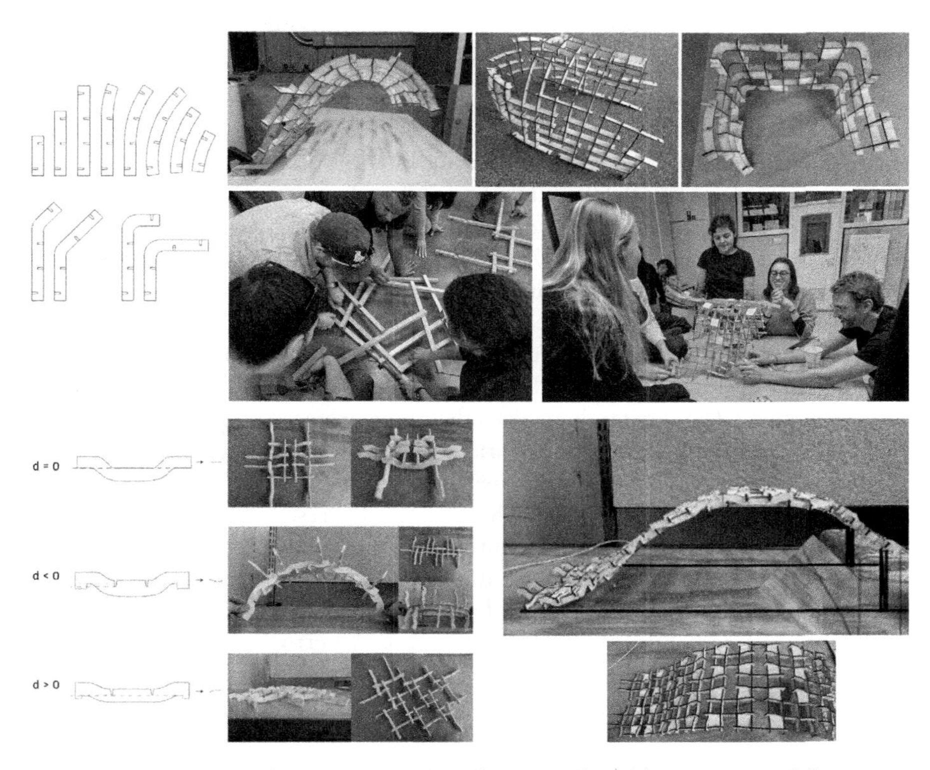

FIGURE 17.4 Laser cut elements can create flat, curved and bent reciprocal frames.

students could see how even miniscule slippage at each joint can accumulate to be a noticeable sag. They could squeeze or push the model to see how forces could distort the form and immediately see the results of adding tension cables and supporting.

Effective physical modeling requires several aspects to come together: defining an appropriate scope and scale for the task; finding appropriate materials, tools and workspace; arranging time for appropriate skill training. In a multi-valent design/build project, this can require a lot of resourcefulness. Collaborative design, software training, woodworking and digital fabrication require different spaces, equipment and support. None of these fits within a typical university structure. So, we relied on a big network of allies for financial, advising and teaching support. While we had arranged woodshop and CNC support for the students, we had not anticipated the need to craft fabric panels for wind and rain protection, nor had we scheduled time to digitally simulate these forces. In a similar way, we could not adequately address outdoor stage acoustics in an 11-week quarter focused on structures. While we would prefer to focus only on timber structures and constructability, weather protection is crucial for wood durability and user comfort.

Assignments for a Mixed Audience

Students from differing majors and levels bring different expectations for course work and evaluation. Those from engineering and construction management expect clear right or wrong quantitative answers, while longer complex problems fit the architecture students' expectations. OSU students expect regular assignment or quiz points, UO Architecture students take studios on a Pass/Fail basis. So, while given similar assignments, tasks and evaluation is tailored to the audience and project needs. Engineering students used to being given a fixed form to analyze were prompted to identify successful structural and construction approaches to inform the project from the bottom-up. Instructors evaluate individual work-in-progress reports on engagement and understanding. Periodic team project submissions are assessed on architectural design, structural efficiency, digital studies, construction detailing and material appropriateness.

Graduate students further enrich the class by researching and presenting a topic of interest, sometimes interviewing an industry expert. Undergraduates often appreciate their ability to summarize complex issues in clear language. Remote lecturers have included Tom S. Chung of Leers Weinzapfel, Lucas Epp of Structurecraft and Alberto Pugnale of University of Melbourne.

Analytics

For structural understanding, students are taught Karamba, a Grasshopper plug-in, to compare alternatives and identify areas of weakness. Engineering students with other expertise will export the structural form and perform that analysis in software such as SAP 2000. While this breaks the seamless interaction between parametric adjustment and analytical results, it supports broader engagement.

Logistical Challenges

Every design/build project requires overcoming unexpected challenges. Our "crises" were excellent learning opportunities. In Fall 2023, as the consolidated class design of an arched pavilion appeared to be ready for CNC cutting, structural analysis showed that wind uplift would be problematic. After initial despair, the team redesigned notch connections into more secure mortise and tenon joints, creating new CNC cutting files that evening. In the next days, students proposed options to make the surface more permeable to wind and planned how additional concrete piers could anchor the fabric. Similarly, the class responded quickly in finding alternatives for CNC machining and the installation location when barriers emerged.

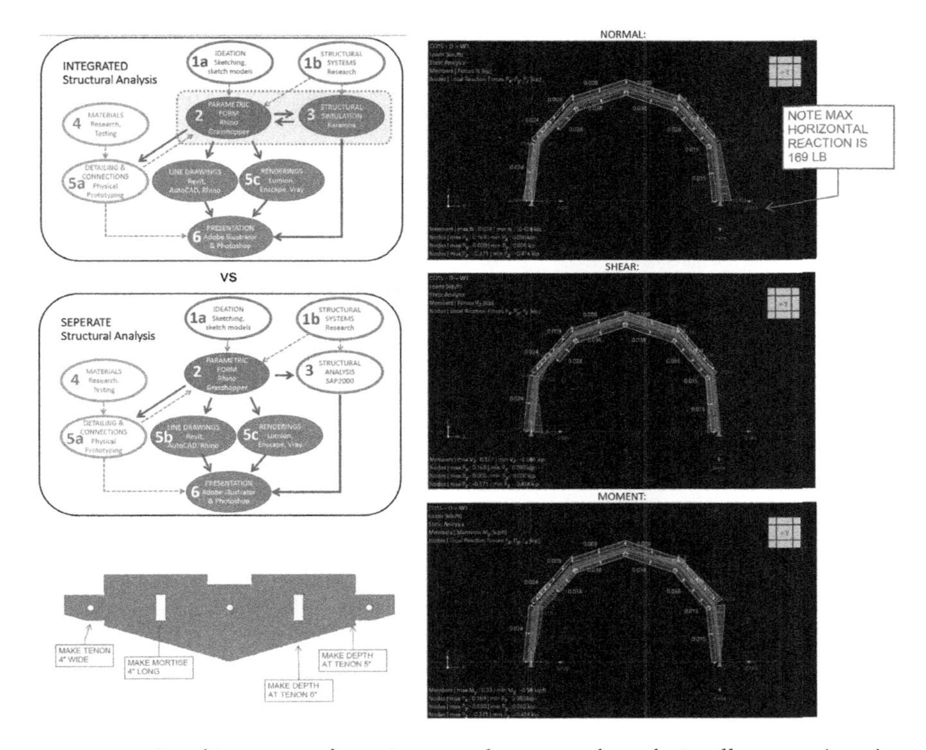

FIGURE 17.5 Breaking away from integrated structural analysis allows engineering students to contribute to the project more fully.

Source: Images at right and bottom left by Nick Thielsen, 2023

Required Readings

Readings evolve with the project focus and include research papers, industry publications and case studies.

Fast & Epp Calculator. www.fastepp.com/concept-lab/apps/calculator/. Member sizing for concrete, wood and steel.

Larsen, Olga Popovic. *Reciprocal Frame Architecture*. Architectural Press, 2008, Chapters 1–2, 4–5. Introduction to the history, geometry and structural properties.

Woodworks (http://woodworks.org) and Timber Development UK (https://timberdevelopment.uk). Timber case studies, awards, product information.

Recommended Readings

Bavarel, Oliver and Alberto Pugnale. "Reciprocal Systems Based on Planar Elements: Morphology and Design Explorations." *Network Nexus Journal* (Turin) (2014) 16: 179–189. DOI: 10.1007/s00004–014–0184-x.

InFuture Wood coordinated Design for Deconstruction and Reuse research efforts across 23 European universities and organizations in 2019–2022. Its publications explain practices for deconstructing buildings, testing, regrading and reusing wood. See

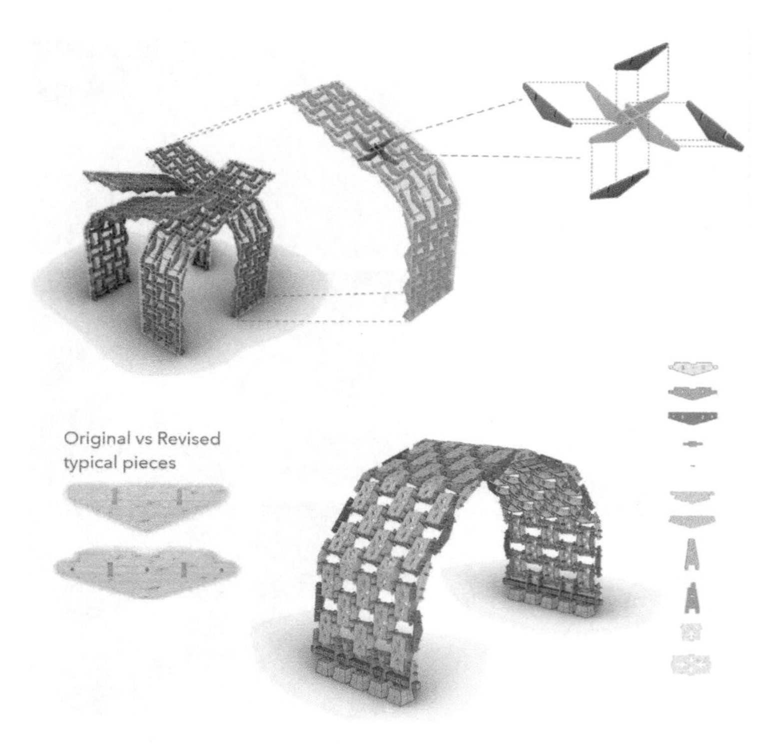

FIGURE 17.6 Original pavilion design (left) with notched module inspired by KKAA[6] and final arch with mortise and tenon joints (right).

Sandberg et al.'s 2022 Summary report *InFutUReWood – Innovative Design for the Future – Use and Reuse of Wood (Building) Components* (www.diva-portal.org/smash/record.jsf?pid=diva2%3A1711749&dswid=3358) and Cristescu's 2020 *Design for Deconstruction and Reuse of Timber Structures – State of the Art Review* (www.infuturewood.info/wp-content/uploads/2021/02/InFutURe-Wood-Report-D2.1f.pdf).
Zwerger, Klaus. *Wood and Wood Joints: Building Traditions of Europe, Japan and China*, Birkhauser. 3rd edition. 2023. Wood connections in cultural contexts.

Outreach Opportunities

We have shared the Timber Tectonics story with academic, professional and public communities in the US, Portugal[7] and Australia.[8] In addressing architecture professors,[9] we explained our methods for supporting student teams and approaches for overcoming challenges of remote partnerships. The importance of communication skills and quick scheming with both physical and digital models could be useful in both education and practice.

Students presented to a professional audience at the 2019 Oregon Design Conference, where architects were interested in how material knowledge and digital skills could be taught in conjunction with teamwork. We showed our

FIGURE 17.7 Construction process shows plywood components that form modular sections, which are shored and connected into an arch.

approach of using reciprocal frames for working with recycled wood to mass timber professionals[10] and educators in dialog with practitioners.[11]

Fall 2023 brought the chance to create a project for the City of Salem, OR with input from city staff through the University of Oregon's Sustainable City Year program[12] that matches classes to civic clients. A big village of support came together to support its realization: Roseburg Forest Products, Tallwood Design Institute, UO College of Design, OSU College of Forestry, and Corvallis CNC Woodcutters provided materials, time, effort and space for our students to realize their design. After building and dismantling the arched pavilion at OSU (December 2023, February 2024), we added corner components to create a folded wall for the HOPES conference at UO in April 2024.

Course Assessment

The course provides students hands-on experience with NAAB Program Criteria (PC) 6 Leadership and Collaboration and Student Criteria (SC) 4 Technical Knowledge, and secondarily provides awareness of PC 5 Research and Innovation (emerging wood design, analysis and construction technologies).

FIGURE 17.8 Arch components were reused for a folded wall, with the addition of a new corner piece.

Collaboration Findings

Complementing our technical focus, we look at how to improve interdisciplinary collaboration. The biggest challenge is myopia when working with those from other fields. Within teams, students with strong verbal, visualization or analysis skills tend to dominate decision-making. In that context, the ability to share the spotlight is the most valuable in creating fully participatory teams. Individuals must have the patience to learn enough about the partner's skills and knowledge to help them fully contribute their best to the situation. In addition to learning about different abilities, they need to delve into different work patterns and expectations, clarify unfamiliar vocabulary and uncover underlying motivations.

To substantiate our intuitive observations about the interdisciplinary collaboration, we tallied how interoperability within teams correlated with markers of successful collaboration for the *Education Sciences* journal. By reviewing students' beginning and end of the term self-assessments, written reflections and illustrated reports, we could trace the interaction process, engagement and satisfaction of team members. In these documents, we identified indicators of Positive Interdependence, Accountability, Promotive Interaction[13] to rate collaboration quality, as well as looking at Construction of Knowledge across interdisciplinary boundaries. The data showed that teams with successful collaborations generally had smooth data exchange between partners.

Teams with more fragmented interactions can still have strong design results through one or two students who drive the process through strong parametric design and/or analysis skills. As we were concerned about the experiences of the more junior students who appeared to be more passive, we specifically looked at how student backgrounds affected team engagement and effectiveness[14]. Student feedback shows that even when the more junior students appear passive, the less experienced students enjoyed being part of a complex project as they have a lot to learn. We also confirmed[15]findings that teams mixing specialties and seniority are more effective. With partners from the same specialty, competing visions are more likely to clash. When team members have fewer overlapping skills, it is easier for individuals to find complementary roles. More group hierarchy reduces the amount of negotiation or intervention needed for decision-making.

Looking Ahead

Better Use of Reclaimed Materials

We are interested in building with "downstream" wood offcuts reclaimed from manufacturing. While we originally planned to use Freres Lumbers' stacks of Multiple Ply Panel offcuts, it is not well suited for small pieces that experience bending. So, we cut the 2023 pavilion from new 23/32" plywood to expedite the building process and improve construction safety and logistics. When we later cut new pieces to extend the toolkit, we found that even with plywood from the same batch, thickness differences from humidity in storage made it impossible to cut perfectly snug joints for both existing and new components.

More study is needed to find the best kind of assembly techniques for building with variable reclaimed wood. Towards this end, we are partnering with Prof. Rafael Novais Passarelli, who leads timber design for the Building Beyond Borders program at University of Hasselt, Belgium with an emphasis on local and reclaimed materials.[16]

More Accessible Process

We are interested in making it easy and efficient to participate in digital design and structural analysis. While physical prototypes speak to all, it is hard to share complex digital files, so often students rely on less efficient methods.

Structural Issues

While we enjoyed demonstrating the easy construction and flexible reuse of reciprocal frames, we understand its downsides: multiple dependencies for stability can be a problem and joint tolerances can accumulate unacceptably across a network. While popular for experimental pavilions, reciprocal frames are little used in practice because construction risks are magnified with heavier, longer members. To better understand the structural behavior of projects, we need to measure deflection under self-weight for the partially and fully built structures, and use this information to inform a simulation under potential wind and snow loads.

Modular Assembly

We are seeking to understand efficient ways to parse timber structures into easily transported, structurally stable and reusable chunks. We maximized wood-to-wood connections in our reciprocal frame, and found connecting sections via multiple joints simultaneously is inefficient. Comparisons show that metal fasteners provide quicker and more reliable assembly and disassembly than wood joints, as the latter can degrade through weathering and repeated use.[17] While the carbon footprint for mining, manufacturing and transporting metal connectors is much higher, it is ameliorated by the long life of reusable fasteners and the ease of recycling metal.

Conclusion

In summary, an agile and focused course supports success. We adjust our course for continued relevance and engagement by responding to student and professional input. Including emerging research and industry practices keeps it stimulating. Because we cannot robustly address architecture, structures, timber construction and digital collaboration in 11 weeks, on-going editing is crucial. In 2020, we addressed modular housing plus long-span design and found that both is too much. So we make trade-offs. The full-scale construction required removing baseline competency requirements, cutting much course content and organizing around fabrication and assembly logistics. We explain mass timber products and buildings, and then prioritize lighter construction for safe and efficient design/build. While it cannot be comprehensive, our course integrates the diverse requirements and rewards of remote interdisciplinary teamwork. Students meet unexpected challenges by learning in real time and developing a team workflow that prepares them for professional life.

Notes

1 Nastaran Hasani, N. Ahn, A. Yari, L. Ouazzani, M. Riggio and N. Cheng, "Demonstrating Mass Timber for a Circular Economy: An Educational Experience" poster (International Mass Timber Conference, 2023).
2 Olga Popovic Larsen, *Reciprocal Frame Architecture* (Routledge, 2008), ch. 2.
3 Valentina Leoni, J. Rosenthal, Z. H. Jin, T. Parr and W. Searcy, "Five Point Gridshell" (Timber Tectonics project, 2018) https://timbertectonics.com/2017/03/22/5-point-canopy/.
4 John Haddal Mork and S. Hillersøy Dyvik, "Gridshell Manual" (MArch thesis, Norwegian University of Science and Technology, 2015).

5 E. Alampi, N. Cheng and M. Riggio, *Timber Tectonics: Building for the Circular Economy* (University of Oregon Sustainable Cities Year Report, 2024), https://scholarsbank.uoregon.edu/xmlui/handle/1794/29262.
6 Kengo Kuma and Associates, "Kyushu Geibun Kan Museum (Annex 2)," https://kkaa.co.jp/en/project/kyushu-geibun-kan-museum-annex-2/.
7 Mariapaola Riggio and N. Cheng, "Timber Tectonics in the Digital Age: Interdisciplinary Learning for Data-Driven Wood Architecture," in *Structures and Architecture: Bridging the Gap and Crossing Borders* (Routledge, 2019).
8 N. Cheng and M. Riggio, "Learning Timber Tectonics Through Digital Collaboration," paper and exhibited video, Computer Aided Architectural Design and Research in Asia proceedings (CAADRIA, 2022), https://doi.org/10.52842/conf.caadria.2022.2.345.
9 M. Riggio and N. Cheng (2021) "Computation and Learning Partnerships: Lessons from a Wood AEC Partnership," *Education Sciences* (MDPI, 2021), 11(3), 124, https://doi.org/10.3390/educsci11030124.
10 Nastaran Hasani, N. Ahn, A. Yari, L. Ouazzani, M. Riggio and N. Cheng, "Demonstrating Mass Timber for a Circular Economy: An Educational Experience" poster (International Mass Timber Conference, 2023).
11 Nancy Cheng, "Shelter from Salvage: Building for the Circular Economy," American Institute of Architecture / Association for Collegiate Schools of Architecture Intersections Conference, Amherst, MA, Oct. 19–22, 2023 (ACSA, 2023).
12 E. Alampi, N. Cheng and M. Riggio, *Timber Tectonics: Building for the Circular Economy* (University of Oregon Sustainable Cities Year Report, 2024), https://scholarsbank.uoregon.edu/xmlui/handle/1794/29262.
13 David W. Johnson and R. T. Johnson, "Cooperative Learning and Social Interdependence Theory," in L. Tindale et. al., Eds., *Theory and Research on Small Groups* (New York: Plenum Press, 1998), 4 (9–36).
14 Nancy Cheng, "Shelter from Salvage: Building for the Circular Economy," American Institute of Architecture / Association for Collegiate Schools of Architecture Intersections Conference, Amherst, MA, Oct. 19–22, 2023 (ACSA, 2023).
15 Richard Tucker and C. Reynolds, "The Impact of Teaching Models, Group Structures and Assessment Modes on Cooperative Learning in the Student Design Studio," *Journal for Education in the Built Environment*, 1(2) (2006), 39–56.
16 Nancy Cheng, R. Novais-Passarelli and M. Riggio, "Timber Structures for Circularity: Reinterpreting Lessons from the Past," International Association of Spatial Structures Symposium, Zurich (IASS 2024).
17 Leonardo Enzo Pozzo, *Design for Disassembly with Structural Timber Connections* (TU Delft 2020), http://resolver.tudelft.nl/uuid:c8b85050-cd11-45eb-b3ac-c3624259654f; Lise-Mericke Ottenhaus et al., "Design for Adaptability, Disassembly and Reuse," in *Construction and Building Materials* (Elsevier, 2023), 400.

PART V

Urban Scale

18

AN INDUSTRIAL-URBAN SYNTHESIS

Planning Education for a Carbon Neutral Future

Craig Brandt

Course Overview

Course Name

Upper Level Graduate Studio

University / Location

University of Notre Dame / South Bend, Indiana, USA

Targeted Student Level

- Graduate
  - 2nd year, Master of Architecture, 2-year program
  - 3rd year, Master of Architecture, 3-year program

Course Learning Objectives

In this course, students will:

- Explore how the design of a physical experience of place can influence critical urban issues that affect wellbeing such as cultural identity, integration with nature and a sense of community while solving practical urban planning and infrastructure planning issues.
- Learn how preservation and revitalization of buildings and natural resources can play a role in a new urban vision and enhance sustainability and resilience.

DOI: 10.4324/9781032692562-24

- Through the process of design synthesis, explore traditional forms and structures of urban planning, while considering modern processes and forms of industry and in the process enable a symbiosis of industry and public life.

Course Learning Outcomes

Successful completion of this course will enable students to:

- Possess an understanding of potential approaches to regenerative urban projects which adopt planning practices and land use regulations that encourage carbon neutral design such as compact transport, walkability and flexible facility design.
- Demonstrate an understanding of the various bio-physical, historical, economic and social-cultural layers of the city, and work with these to form a synthesis with their regenerative designs.
- Demonstrate abilities in teamwork and time management for group and individual work, as well as development of a long-term project from large scale to medium and small scale.

Course Methodology

Philosophy

As an educator who teaches concepts of regenerative and adaptive design, my teaching philosophy centers on providing urban projects that allow students to adapt to various modern day challenges using ideas and research from a broad spectrum of historical precedents and in-depth observation. With a large population of the world residing in urban areas, the role of education to study its impact on climate is critical, especially at the planning level. Vehicular urban and transport planning during the modern movement has led to high levels of pollution, heat island effects, and lack of green space and physical activity because of changes to the structure of industry and living. How can new considerations for urban transport and use planning, industrial rethinking, and an overall more synergistic approach improve the carbon footprint of our cities? One central idea looking at urban transport performance in this studio is to create more integrated, hybrid-industrial use typologies that can be adjacent to civic and residential uses, promoting nodes for future transportation and public gathering for retail and cultural activities. Chicago's industrial past, present and future is a fitting place to explore the idea of industrial-based hybrid typologies. As formally a leader in establishing planned manufacturing districts (PMD) in the 1980s, Chicago is now starting to redact these PMDs in areas such as Lincoln Yards for large private development. It is critical to explore alternate paths for the relationship between industry, policy and place for future generations.

Definitions

Carbon neutrality – Balancing or reducing the production of carbon dioxide over a described time period.

District use diversity – A quantitative ratio to track a subjectively distributed mix of use types.

Urban cluster – Dense cores of legible blocks with adjacent densely settled surrounding areas lacking strict boundaries.

Walkability – Accessibility of amenities and resources by foot, most often defined as a 15-minute walk.

Implementation

Commencing as a group, the students were tasked to create a 128-acre "15-minute industrial center" through an illustrative regenerative master plan that blended different intensities and scales of industrial block uses planned within a compact street network, while also considering integrated landscape. This illustrative regenerative master plan would be further developed with clusters of specific hybrid programs as an urban use layer, and lastly, designing individual development structures in each cluster acting as incubators for growth. The implementation consisted of four parts:

Part 1 Analysis: consisting of a group site visit, site inventory, urban analysis and the mapping of existing conditions to create a base plan. Redevelopment goals and precedent-based ideas would be marked up upon this base plan.

Part 2 Creation of Illustrative Urban Regenerative Plan: Based on urban hierarchies and confluences, including basic programmatic districting, critique #1.

Part 3 Development of Cluster Zones: Pairs of students design development cluster zones and reintegrate back into the master plan, critique #2.

Part 4 Facility Development: Individual students design new hybrid facilities and site plans while reintegrating back into the master plan, critique #3.

Part 1 Analysis

Inventory

The project commenced with a site visit to Chicago, observing vast unused industrial land along the river and adjacent to the thriving community of Pilsen. The site, located within one of Chicago's PMD zones, was inventoried alongside urban conditions within one square mile. The inventory included several forms of data: cultural and physical histories, use type maps, Chicago River development data, and a profile of the existing urban transport network, focusing on the history of former and current industries and commercial developments, river regulations, edge studies and community input reports.

Urban Analysis

Starting with transport as a primary design parameter, biking and walking distances to public transportation were charted on scaled maps. Half-mile walkability sheds were overlaid on the base plan to determine access to programmatic resources, with an existing use analysis of the surrounding Pilsen community revealing growth and density patterns documented through color-coded figure/ground illustrations. These findings were composed with data from the site inventory to create a booklet for presenting findings and rationale to the jurors. This booklet also helped establish a consensus of goals and directions within the student group for application of urban redevelopment schemes.

Precedents

Given the limited discussion and construction of industrial-type hybrid typologies in contemporary urban design programs and organizations, a diverse range of precedents was examined, varying in scale and location, all of which featured industry as a primary use within urban settings. The Pullman factory town serves as a historical example of an industrial complex, meticulously studied for its scale, use types, forms and urban layout, reflecting the region's industrial heritage. Furthermore, industrial adaptive reuse models in the United States and abroad were explored, showcasing mixed-use interventions that capitalize on the unique opportunities presented by repurposing industrial buildings for non-industrial and mixed-use purposes. Examples from urban plans in East Portland, Oregon, and the Third Ward of Milwaukee, Wisconsin, illustrate unplanned communities characterized by a mix of industrial and non-industrial uses, often featuring finely grained block structures and distinctive vertical buildings. Newmarket Square in Boston, Massachusetts, emerged as a prominent precedent, notable for its well-planned development approach that integrates unique typologies across a range of legible and coded industrial types. Finally, the study encompassed precedents involving urban approaches and sustainable strategies along riverfronts, notably along the Chicago River, which has seen full development primarily in downtown areas. Insights gleaned from these precedents, including travel distances, block densities and open space ratios, serve as a foundational basis for future comparisons.

Base Plan and Site Analysis Diagram

A base plan at 1′ = 100′ was completed, outlining existing major streets as boundaries and minor streets as planning grid lines, transportation stops and river edges. As a part of developing the base plan, the students would have to determine which existing urban elements to preserve, such as the river slips and industrial buildings including the iconic Fisk Power Plant. They were also encouraged to leverage existing cultural layers, and thus, a connection to the existing Pilsen mural walk was additionally included within the base plan. Locating the existing urban transport links showed the bias of transport to the east side of site to extend a connective network westward. These goals for reuse and connection would be illustrated over the base plan in a single "Site Analysis Diagram."

Part 2 Creation of Illustrative Urban Regenerative Plan

After identifying opportunities for reuse and connection, the student group must apply their analysis findings to make informed decisions about new urban interventions. The primary focus is on extending the existing Pilsen neighborhood from the north of Cermak, a bustling industrial street, southward over industrial property to the Chicago River, while introducing new developments from east to west, aligned with a connective street structure to facilitate efficient movement toward the transit centers on the east side of the site. Commencing with a common base plan, students engaged in a one-day charrette, individually testing various urban schemes to generate several consolidated plan options, eventually converging on a single legible urban plan. Once established, public-oriented uses and urbanistic features were strategically located at perceived points of confluence to maximize the plan's regenerative potential. The base plan, with its adjacent street grid lines, served as a foundation for plotting a series of public-oriented nodes along the river slips and Fisk Power Plant. Subsequently, possible development centers were proposed and distributed within the plan using the three walkability sheds, each 0.5

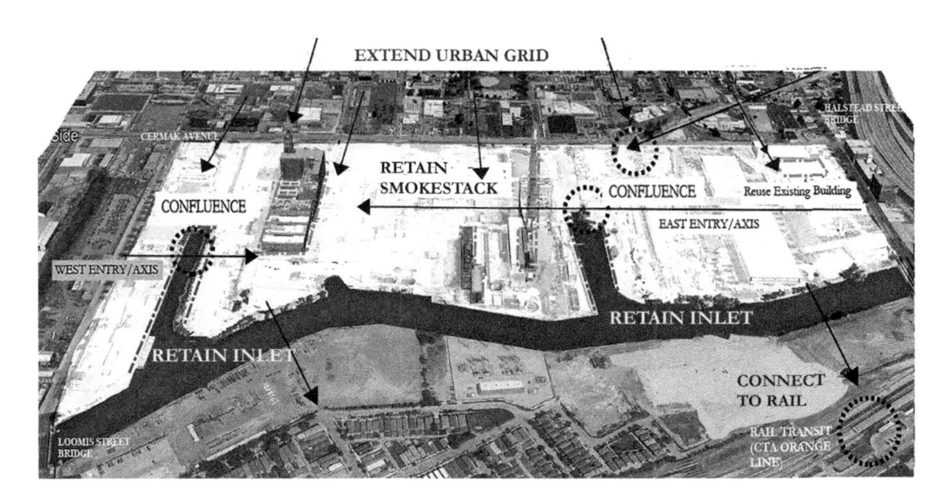

FIGURE 18.1 The site analysis graphical process highlighted confluences between landmarks, resources and possible extensions of street structure.

miles in radius, as demarcations of urban sections. The remaining parcels were designated for private developments with mixed-use to industrial uses. The development followed a compact block-type structure based on a typical Chicago block, with variations depending on orientation and location.

The central node of the plan is located around the Fisk Power Plant whose tall and iconic smokestack can be seen from surrounding areas as a sculptural feature of a new main public plaza. From this central node, an adjacent existing river inlet was extended north to draw public from Cermak Street along a promenade. Flanking from this central node, an east-west main street interconnected additional promenades to increase connectivity from Cermak Street to the Chicago River.

Enhancements for additional urban transport connectivity would be the reinstallation of a street bridge at the river that was once removed, connecting Bridgeport to the South and reducing the number of dead-end streets at the river. A special pedestrian bridge leading to the CTA Orange Line train station was proposed to allow pedestrians a more direct route from the new redevelopment zone.

After the development of the urban plan, program envisioning involved the distribution and clustering of use types. The metric of *district use diversity* would include an allotment for industrial uses, along with living, commerce and recreation, in a manner that distributes proportionately within and across the walkability sheds. Finally, testing of program locations and adjacencies was completed through an additional charrette process, during which students were divided into pairs to envision how locating programs would create interesting

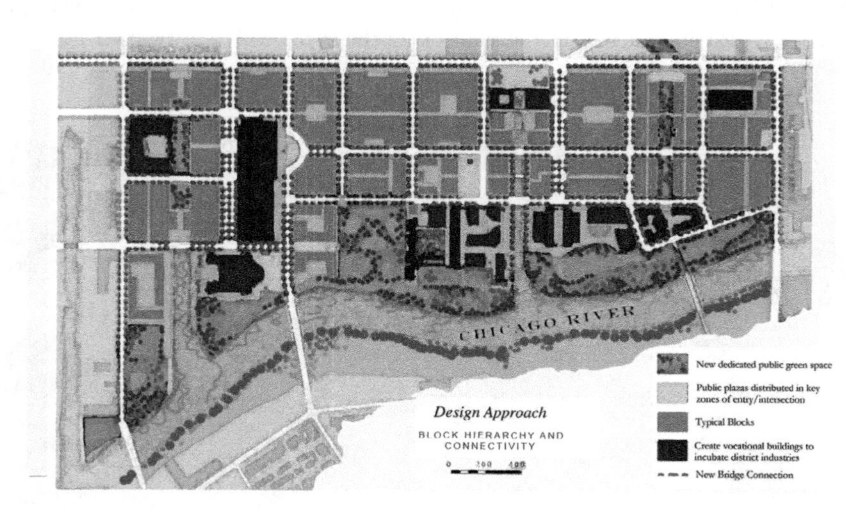

FIGURE 18.2 Urban form at an early stage shows integration of street structure, resources and likely locations for higher densities of public gathering, becoming future regenerative centers. These sites distinguish themselves from the typical blocks. New bridge connections proposed at the Chicago River and greenway connections to Pilsen at the north.

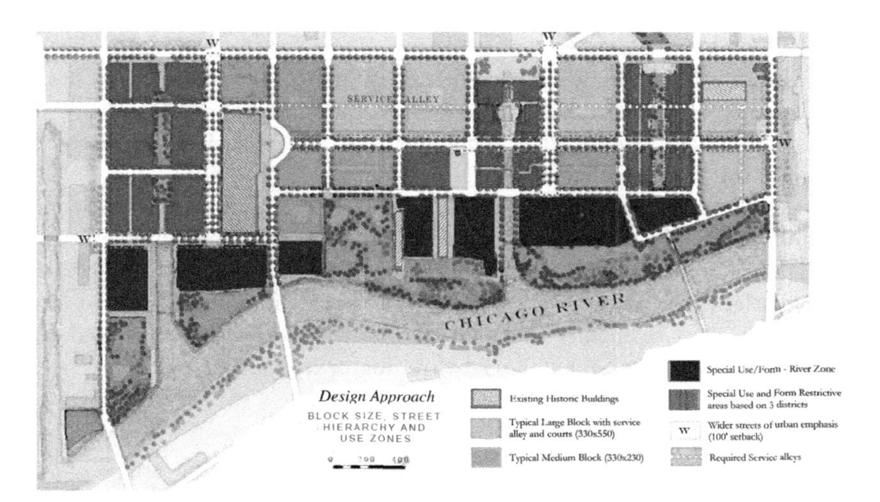

FIGURE 18.3 Hierarchies for streets and blocks distribute flexible uses versus more prescriptive uses.

public experiences and movement within the new development area. A presentation to the faculty included the initial analysis and proposed regenerative urban plan for commentary before the next phase of work.

Part 3 Development of Cluster Zones

Considering not only the physical aspects of urban planning, but also the programmatic aspects, allows students to explore a range of communal potentials that illustrate future growth and vibrancy, as well as explore a wider range of building typologies. After receiving critiques on their initial urban regenerative plan, the students were divided into subgroups and assigned to assess similar urban objectives of the master plan at a finer scale of 6–8 blocks or approximately 40 acres. At this scale, they were tasked with considering ideas to define legible industry clusters consisting of specialized complexes for production, adjacent to a diverse mix of residential, recreation and shopping areas. Each area would leverage existing features and strengthen the urban nodes developed from the urban plan. The midterm presentation models and renderings highlighted a thematic layer of these types along with the locations of special facades, walking spaces, new bridges and other connective features as part of an integrated urban experience. What each zone has in common is some kind of "industry to commerce" or "industry to public building" chain structure as an anchored linear progression.

The western group developed a campus type cluster form that would utilize a large-scale industrial building as a northern anchor and circulate along a river slip inlet promenade to a multi-use public facility located at the opposite south end at the Chicago River. The campus type was used as a way of leveraging the intimacy of the river slip inlet.

The central group would similarly propose a river-slip promenade that circulates from a public building complex at the north to an urban farm development adapted from the existing historic Fisk Power Plant. The central section has the largest contiguous set of regular blocks, and as such it was deemed the best place for higher scaled industrial blocks.

The east group proposed a more residential based promenade to cluster work-live and multi-use buildings along the extension of the existing Pilsen mural green-walk. Like the other groups, this development extends south to two anchor developments at the interface with the Chicago River which is lined with public parkland.

A presentation to faculty included the initial analysis and proposed regenerative urban plan for commentary before the next phase of work.

Part 4 Facility Development

The purpose of the next assignment is to enable students to develop new hybrid structures to serve as anchor developments for their respective clusters. These hybrid buildings are intended to mirror the role of civic buildings in traditional urban plans but with a modern twist, emphasizing how production can act as a vibrant incubator of activity within urban areas when planned using contemporary principles and flexible structures. As integral components of the cluster zone, these buildings are designed to seamlessly integrate with the surrounding neighborhoods, fostering community gatherings and enhancing accessibility to main transport networks. Notably, the existing CTA Orange Line station lacks any memorable urban spaces that encourage walking or biking.

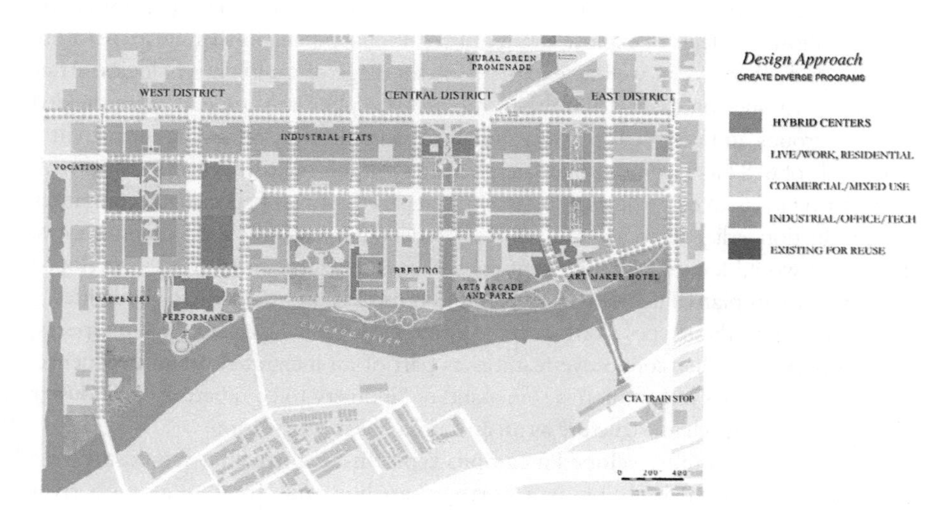

FIGURE 18.4 Three cluster zones (based on three walkability sheds) establish industry types and propose incubator development parcels within each cluster. Each cluster zone has connectivity to the river and to the existing urban fabric.

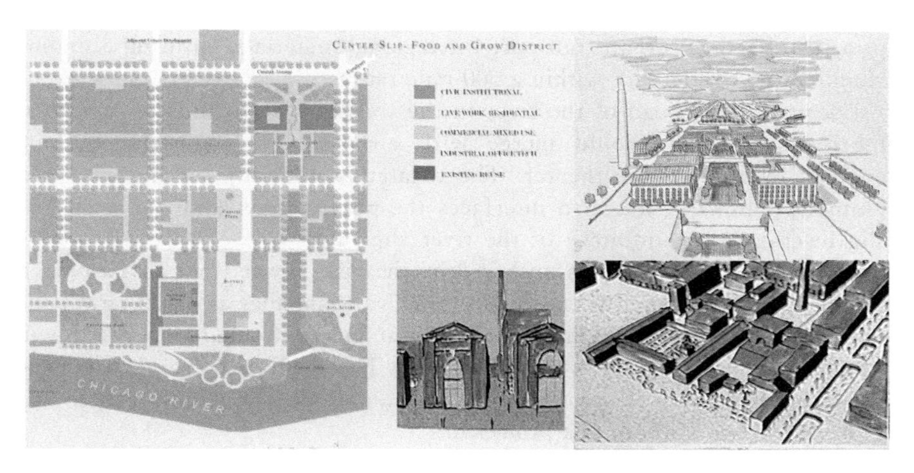

FIGURE 18.5 Character and use studies for the Center District combine urban farming, brewing and community vocational uses along a promenade of industry/retail buildings.

Source: Illustrative sketches to assess scale and character by Anna Drechsler and Braiden Green

Drawing from a range of precedents discussed in class, the organization of the hybrid building type can encompass various programs and uses, inspired by examples from Chicago. For instance, downtown office buildings like the Monadnock Building feature ground floors hosting an indoor promenade of shops, including shoe repair and other service-oriented shops, with offices situated above. The Thompson Center, before being sold to developers, boasted a unique mix of civic and office uses, complemented by a food court within an impressive atrium space. Additionally, more industrial-centric buildings in Chicago, such as the Metropolitan District, focus on food and beverage production, offering brewing facilities alongside a distillery, coffee roaster and a public event/bar facility situated along the Chicago River. The Brewery type provides a unique experiential journey where production processes can be observed, promoting the concept of "farm to table" and "product to store" narratives, enriching the public urban experience.

In the realm of housing hybrids, the 21C hotel chain incorporates an art gallery as a public feature within its transient residential use type.

Once students have programmed and sized their facilities, they focus on developing the experiential and functional aspects of the block plans, including the integration of public entries and vehicular access points for loading and parking. Massing studies are undertaken to refine the urban relationships within the facility, emphasizing scale, legibility and dialogue between functions both internally and within the surrounding urban context. Elements such as development wings, production courtyards and market alleys are carefully considered to unite function with urban legibility. Finally, materiality choices prioritize compact urban transport for regional materials, ensuring economic viability and availability within a 500-mile radius.

Located at the head of the river in the western "arts campus," a scenery fabrication facility would merge with a large-scale industrial building enabled by a multi-use theater. The theater is not only serving as a programmatic linkage, but also interfaces the public street and the west river slip leveraging the intimacy of the river slip inlet.

The central area cluster is anchored by the Fisk Power Plant redevelopment which would repurpose the existing historic brick buildings as urban farm structures used for growing and markets. The urban farm is fronted to the north with a brewery complex that faces the urban plaza. The north anchor of the river inlet is a mixed building with a community center and trade school with retail located between the two anchors as a promenade.

The east group facilities anchor the river edge for a more residential based promenade with a hotel-museum-studio hybrid facility and an office/art gallery hybrid facility with a glazed public galleria. These developments draw pedestrians to the pedestrian bridge which connects to the CTA transit station.

The unique building programs described highlight the inspired potential that future generations of urban designers and architects may propose to help bring our urban areas back to a more symbiotic coexistence.

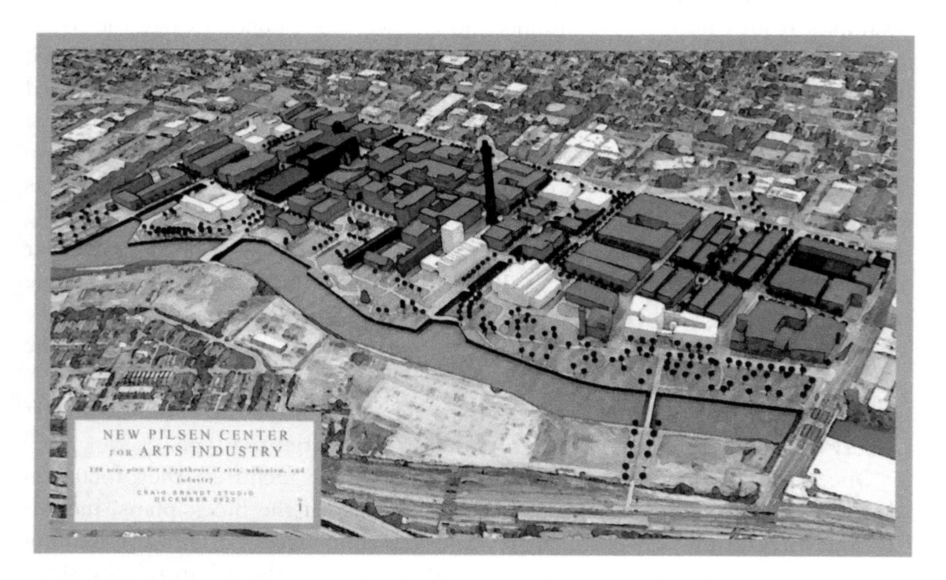

FIGURE 18.6 Illustrative rendering of final development: hybrid developments (modeled in white) create civic terminations at urban confluences. Typical blocks are represented as small and medium sized structures that can be combined for larger industrial uses.

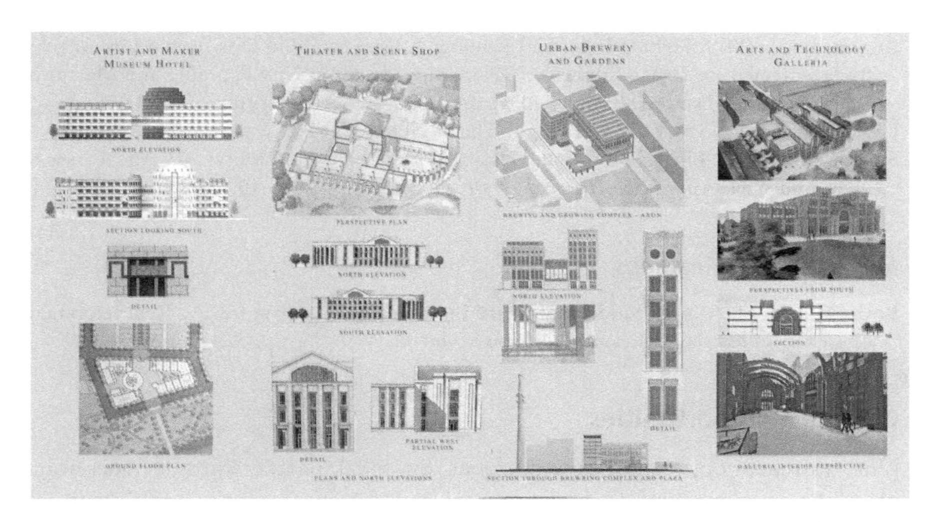

FIGURE 18.7 Final projects integrate industrial uses into moderately scaled buildings with urban responsive features.

Source: Illustrative sketches by (from left to right) Alessandra Giannasca, Dylan Rumsey, Anna Drechsler and Monica Medina

Analytics

Data Analytics were utilized in this project for quantitative measurements of spatial characteristics, connectivity and density, which are aspects of urban planning, as opposed to materials research and energy savings. The missing link between "green city" design and overall performative design for human health and welfare is the lack of urban metrics such as public access, walkability and other facets of efficient transport. To conduct this analysis, Autodesk CAD and Forms software, SketchUp and Acrobat were employed in conjunction with Google Earth. These tools were instrumental in quantifying land use metrics, allowing for a comparison of existing and planned development quantities, as well as development ratios between the three incubator districts. Notably, a critical development center of the project site was located approximately 0.37 miles from the transit stop, with the addition of the pedestrian bridge significantly enhancing accessibility. Overall, the students achieved a 35% industrial program mix, 20% public landscape allocation and a density ratio of 28.1. Comparatively, these metrics align with certain aspects of their precedent studies, serving as valuable learning tools for understanding how precedent informs design outcomes. In future projects along the river, comparisons between each project will be conducted using more robust parametric tools.

Required Readings

Boston Planning and Development Agency. 2019. PLAN: Newmarket The 21st Century Economy Initiative.

Brown, Sarah. 2019. Hybrid-Industrial Zoning: A Case Study in Downtown Los Angeles By Department of Urban Studies and Planning; Master in City Planning and Master of Science in Real Estate Development at the Massachusetts Institute of Technology.
Gibberd, Fredrick. 1959. *Town Design*, The Architectural Press, pp. 67–262.
Stangl, Paul. 2015. Block Size-Based Measures of Street Connectivity: A Critical Assessment and New Approach. *Urban Design International*. DOI: 10.1057/udi.2013.36

Recommended Readings

Love, Tim. 2017. A New Model of Hybrid Building as a Catalyst for the Redevelopment of Urban Industrial Districts. *Built Environment*, 43(1) (Spring 2017), pp. 44–57.

Outreach Opportunities

Our outreach opportunities will focus on local partnerships to encourage visioning sessions and forums, to see how compact travel and integrated development can take shape across several potential areas of the city. Our potential partners include the Chicago Architecture Center, the Chicago Department of Transportation, and the Department of Planning. The ability to blend both industry and culture into this integration will yield partners with diverse groups such as the Pilsen Community Outreach and Urban Growers Collective. Finally in partnership with the Friends of the Chicago River we can look to promote more river-centric design discussions to enliven this much underutilized urban resource.

Course Assessment

Critics for assessing the midterm and final reviews came from a range of geographies including Spain, South Bend, Indiana, New Haven, Connecticut, and a representative of the City of Chicago Department of Buildings. On the whole, critics found the range and diversity of the projects inspired and the range of urban analysis commendable. It was thought that the project had a strongly "internal" generative approach which may not provide true connectivity past the edges of the site boundaries.

When considering weighting of required NAAB criteria for this class, the structure and context of this studio allows for students to develop skills most notably in design, design synthesis, ecological knowledge and responsibility, and leadership and collaboration. The proposition that industrial-urban uses are considered potential for being considered "civic" incubators, although respected and attempted by the class, was not consistently explored. Half of the students proposed more commercial and institutional uses for their hybrid developments. As the future of office and industry as places of business develop post-COVID 19 pandemic, it will be of interest to reflect on the proportions of work-live proposed in these studio projects.

Looking Ahead

Future class projects along the Chicago River are in process and will continue to be used as study specimens for potential regeneration of PMD districts. Besides considering new ways of achieving hybrid design approaches, data will be compared between each project using parametric tools that can illuminate relationships between these urban approaches and the current urban condition in other Planned Manufacturing Districts such as Lincoln Yards.

Acknowledgements

All plans and diagrams are attributed to the class, Anna Drechsler, Alessandra Giannasca, Braiden Green, Monica Medina, Dylan Rumsey and Nicolas Sabogal Guacheta, under the instruction of Professor Craig Brandt. Plan and Model images were edited for publication by Craig Brandt.

19

COMPUTATIONAL URBAN DESIGN

A Simulation and Data-driven Approach to Designing a Sustainable Built Environment

Timur Dogan and Yang Yang

Course Overview

Course Name

A Simulation and Data-driven Approach to Designing Sustainable Cities

Universities / Locations

- Cornell University / Ithaca, NY, USA
- Thomas Jefferson University / Philadelphia, Pennsylvania, USA

Targeted Student Level

- Undergraduate
  - 4th or 5th year, Bachelor of Architecture, 5-year program (Cornell University)
- Graduate
  - 2nd or 3rd year, Master of Architecture, 3.5-year program (Cornell University)
  - 1st or 2nd year, Master of Science Advanced Architectural Design, 2-year program with optional 1.5-year advanced standing (Cornell University)
  - 2nd year, Master of Urban Design, 2-year program (Jefferson University)

Course Learning Objectives

In this course, students will:

- Learn to build data-rich models of urban sites and derive insights that inform strategic urban design decisions.

DOI: 10.4324/9781032692562-25

- Gain proficiency in creating parametric urban design models, enabling the systematic, rigorous exploration of design variations.
- Master using advanced building performance and mobility modeling tools to quantitatively assess urban designs.
- Apply systems thinking to address urban design challenges, utilizing multi-objective optimization techniques to refine design schemes effectively.

Course Learning Outcomes

Successful completion of this course will enable students to:

- Effectively analyze complex urban systems and develop and support sustainable design strategies using a data-driven and simulation-based approach.
- Develop a profound understanding of the connections between the morphology of the built environment and sustainability outcomes.
- Systematically explore design variations using systems thinking and trade-off analysis to achieve desirable solutions that address multiple sustainability goals.

Course Methodology

Philosophy

Rapid population growth and urbanization are increasing the construction and densification of urban centers globally. By 2060, the UN predicts a construction demand of 250 times that of New York City.[1] This trend is concerning as buildings are responsible for 39% of greenhouse gas emissions,[2] and this figure rises to over 60% when considering mobility-related emissions that can also be attributed to how we design cities. However, the construction and renewal of urban centers present opportunities to mitigate climate change through intelligent design solutions. These solutions aim to enhance energy efficiency, renewable energy use, mobility and quality of life. Future urban habitats must be resource-efficient, resilient, and designed to provide comfortable indoor and outdoor conditions, daylight access, high-quality public spaces and innovative transportation solutions. They should also be capable of handling extreme events, such as pandemics, storms and heat waves. Urban-scale design projects that address such complex environmental and societal constraints necessitate a rigorous, collaborative, cross-disciplinary approach, incorporating quantitative analysis at the critical early stages of design. This studio aims to provide an integrated learning experience that addresses topics of urban sustainability and advances performance evaluation in the design process by closely linking student learning with the authors' research on developing advanced computational urban design tools. Additionally, the course fosters connections with stakeholders and professionals through targeted outreach to design professionals and local planning agencies.

Definitions

Carbon neutrality – We typically use the Annual Site Energy Net-Zero Carbon approach where remaining emissions can be offset with on-site renewable energy. However, limitations of this approach are discussed and the impact of different system boundaries are tested and a monthly load match for renewables is computed to understand the reliance on the grid as virtual energy storage.

Parallel coordinate plot – A visualization tool that allows the representation of multi-dimensional data by plotting each attribute on a separate vertical axis, all parallel to each other, enabling the analysis of relationships and patterns across multiple variables simultaneously.

Pareto Front – A concept in multi-objective optimization that represents a set of solutions where no one objective can be improved without worsening at least one of the other objectives.

Implementation

General

This course has undergone several iterations and evolved in parallel to developing software tools that facilitate data processing, urban-scale simulation of energy,[3] daylight access[4] and active mobility predictions,[5] as well as microclimate and outdoor thermal comfort.[6] To date, the course has been implemented at Cornell University and Thomas Jefferson University, targeting graduate and undergraduate students in Architecture. The educational framework and general methodology consistently motivate students to rethink our urban built environment to respond to societal and environmental challenges by utilizing data to inform design strategies. This chapter showcases examples from work with project sites in New York City, Philadelphia, Toronto and Ithaca. In these examples, students developed projects with a broad range of topics and unique emphasis, depending on the site's context and city. These include exploring mixed-use, densely populated urban typologies to reduce the need for transportation and proposing innovative mobility solutions that promote shared mobility over private vehicles. Additionally, designs investigate density and urban form with respect to daylight access and ventilation, including the design of green areas and outdoor spaces to improve the quality of life and urban microclimates to promote public space use and active mobility, such as walking and cycling. Projects have also incorporated sustainable landscape and infrastructure systems, including rainwater harvesting and water recycling, to enhance sustainability.

Assignments

The studio course unfolds in three distinct stages. In the first stage, teams of 2–3 students delve into research, analysis and diagramming of urban systems relevant to their site (*Assignment 1, A1 – see below*). Simultaneously, students acquire proficiency in advanced simulation and analysis tools to evaluate environmental aspects like daylight, mobility, energy and thermal comfort (*A2*). In the second stage, these teams employ parametric modeling to craft initial masterplans. These plans integrate residential, office and commercial components to transform the site into a hub for urban and social engagement (*A3*). The final stage involves students, individually or in smaller groups, honing in on the specifics of particular buildings or clusters within their master plans (*A4*). This stage confirms the viability of the urban concepts introduced earlier.

A1 Site Analysis and KPI Development

In the first assignment, students learn how to conduct a data-driven analysis of an urban site to identify primary sustainable design challenges. Using open databases and advanced tools, students gather comprehensive data on the site's current conditions, including climate, land use, population, transportation networks, building types and green infrastructure. Table 19.1 summarizes a series of data sets and

TABLE 19.1 Frequently used data sets that are available nationwide. In addition, local data sets from municipal GIS portals or local partners provide valuable site analysis information.

Data set	*Application*
Building footprints, points of interest and street networks	Building footprint geometry is used in the 3D reconstruction of the city. Street networks and points of interest are used in the mobility analysis modules.
Areal Lidar Scan as a 3D point cloud	3D reconstruction of the city, including buildings, terrain, trees and water features. The data is used to create geometric representations used as simulation inputs.
Tax Assessment Data at parcel or building level	Data describe properties and then build on them. Data includes zoning type (i.e., residential, commercial), property value, age, construction type and utility connections. Data is in building characterization for energy modeling.
CBECS and RECS energy consumption survey data	Commercial and Residential Buildings Energy Consumption Survey provide building characteristics information for the estimated 5.9 million U.S. Building characteristics ad will be used as a validation data set.
Public Use Microdata Sample	It contains records about individual people or housing units used to derive demographic information of an area in the Urbano.io mobility model.
Longitudinal Employer-Household Dynamics	The data set provides public-use information on employers and employees. Used to describe home-work commutes in the Urbano.io mobility model.
Federal Highway Administration: National Household Travel Survey	Provides data on travel behavior. It includes daily non-commercial travel by all modes, characteristics of the people traveling, their household and their vehicles.

their applications in the studio. While most data sources are available everywhere in the US, some data access requires collaboration with a municipal government (permit data for building, electrical and plumbing) or a local utility (high-resolution energy use data). This data collection and visualization effort supports the development of a detailed baseline site model that provides a deep understanding of the relevant urban systems associated with the design challenges and allows students to identify neighborhood qualities, challenges and problems as part of the analysis. After an initial data exploration and interpretation, the studio visits the site to validate and augment their analysis by collecting photo impressions and, in some cases, speaking with locals. This exercise aims to reveal the limitations of data-driven site analysis that may portray a biased or incomplete picture.

Subsequently, students define specific sustainability goals for their design project, such as reducing total building energy use or decreasing transportation emissions. These are juxtaposed with considerations of quality of life, economics and equity. Students then associate their design goals with a set of provided key performance indicators (KPIs) to guide their analyses. In addition, students are also encouraged to critique KPIs and, if necessary, expand and complete their set of KPIs. Table 19.2 describes several KPIs that are often used in student projects.

Figure 19.1 (left) shows the site analysis for a project focusing on enhancing site accessibility through active or shared transportation modes to lower emissions. It analyzes urban mobility systems and the need for infrastructure enhancements like sidewalk improvements and bike lane additions. Figure 19.1 (right) analyses Philadelphia's open spaces, exploring the demands and possibilities for developing diverse ecological landscapes across different neighborhoods based on population, amenity densities and vacant lands. Figure 19.2 shows a city-wide analysis of building-integrated solar energy potential and urban building energy simulations for the City of Ithaca to inform urban design strategies that could support Ithaca's Green New Deal,[7] one of the most ambitious decarbonization policies in the US, that challenges building owners, developers and utilities to support large-scale electrification through building retrofitting and new high-performance developments.

A2 Parametric Investigation into KPIs

The second assignment engages students in parametric investigations to examine the relationship between spatial configurations and sustainability metrics. Firstly, students develop a parametric massing model, defining key parameters such as building height, density, orientation, block sizes and program types within constraints like site boundaries, zoning rules and environmental conditions. Then, students develop automated workflows for evaluating KPIs that react to parameter variation, utilizing performance modeling and simulation tools for this analysis. The findings from these

TABLE 19.2 KPIs defined in student projects in the course.

KPI	*Calculation Method*
Sustainability	
Daily Total Travel Distance by Walking/Biking/Vehicle/Public Transit (meter)	Run mobility simulation for the nearby neighborhoods and take the sum of all trips' distance by different modes.
Daylight Access (hours per year)	Number of hours that the analyzed area has direct access to daylight.
Water Recirculation Equivalence (cubic meters)	Comparing the volume of water captured and reused to an equivalent amount of freshwater that would otherwise be required.
Energy Use Intensity of Buildings (KWh per square meter per year)	The annual energy consumption of all buildings relative to their total floor area.
Solar Energy Potential (KWh per square meter per year)	Total amount of solar energy that can be generated per unit area over the course of a year.
Livability	
Outdoor Comfortable Time (hours per year)	Number of hours that the analyzed outdoor space has a comfortable range in the Universal Thermal Climate Index (UTCI) metric (Feels like temperature between 9°C and 26°C).
Accessible Essential Services within 15 Minutes Walking (count)	Average number of essential services and amenities, such as grocery stores and hotels, within 15 minutes' walking distance to all buildings.
Green Space Area Per Capita (square meter)	Total accessible green space area in the surrounding area divided by population.
Job Counts within 15 Minutes Walking (count)	Total number of jobs within 15 minutes' walking distance to all buildings.
View Access to Key Landscape / River (square meter)	Total area of building façade that has unobstructed views of key landscapes in the vicinity, such as Central Park or the adjacent river.
Economics	
Floor Area Ratio	Total building floor area divided by the land area.
Residential Floor Area Per Capita (square meter)	Total residential floor area divided by population.
Length of High-Footfall Block Frontage (meter)	Total length of block frontage experiencing daily pedestrian traffic exceeding a specified threshold, such as 1,000 pedestrians per day.
Required Sidewalk Enhancement (meter)	Total length of sidewalk requiring pavement enhancements, determined by assessing both the estimated pedestrian flow and the existing pavement condition.
Required Bike Lane Addition (meter)	Total length of new bike lanes required to accommodate the majority of biking trips, for example, more than 70%, within a defined area.

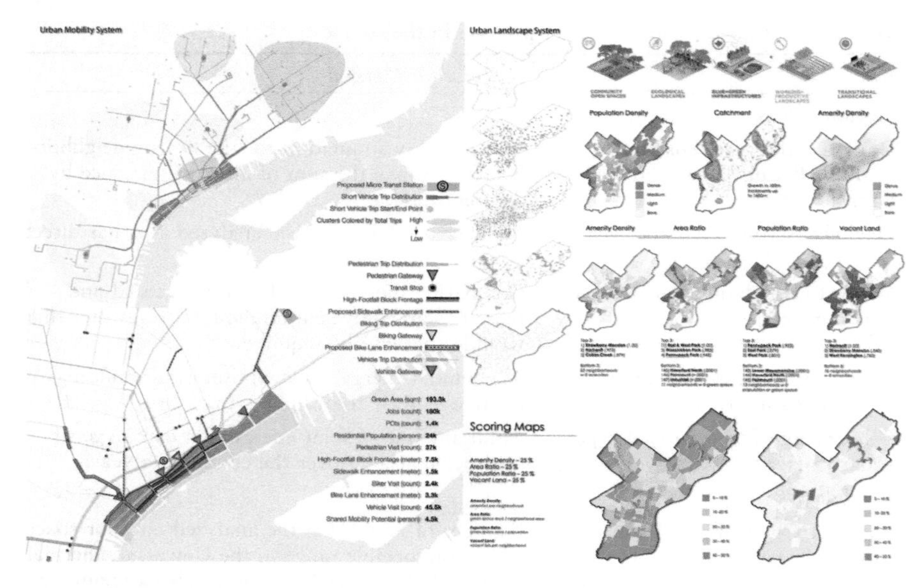

FIGURE 19.1 Left: Urban mobility analysis at the Philadelphia waterfront, illustrating patterns for pedestrian, cycling, vehicle and public transit modalities. Right: Assessment of urban landscape development potentials in Philadelphia neighborhoods, based on factors such as population density, amenities and vacant land.

studies are presented through maps, charts, diagrams and 3D models, revealing insights unattainable through intuitive or qualitative methods alone. For instance, Figure 19.3 depicts a parametric analysis of urban density distributions and street network configurations, examining their effects on traffic demands. The analysis indicates that in a small, constrained urban setting, the average block size – or the density of street intersections – plays a more critical role in influencing the total travel distances for residents' daily commutes than density distribution. Figure 19.4 demonstrates an innovative approach to visualizing pedestrian flows in response to modifications in building layout. The analysis reveals that even minor adjustments, such as altering the position or orientation of a few buildings, can significantly impact aggregate pedestrian flows on the site. Figure 19.5 examines the relationship between daylight potential and the massing form of urban typologies, demonstrating that well-performing designs can significantly increase built density without compromising daylight performance. Figure 19.6 illustrates how urban projects designed for optimal solar access and wind sheltering can extend the thermally comfortable time by approximately four months. Together, these examples highlight how performance analysis can substantiate the benefits of effective urban design and demonstrate a value proposition that designers should leverage in their work.

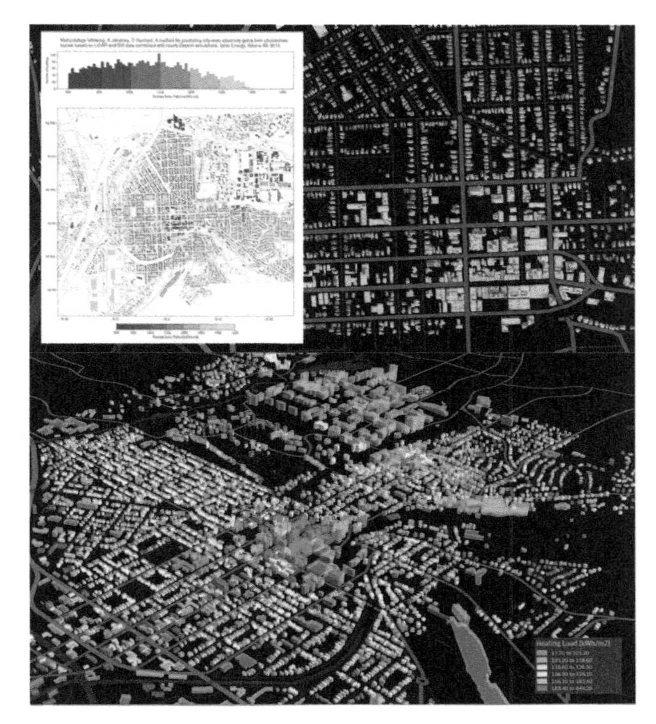

FIGURE 19.2 An analysis of Ithaca's city-wide, renewable energy potential through building-integrated photovoltaics and building-related heating energy demand predicted with an urban building energy model.

A3 Design Exploration, Optimization and Trade-offs

The third assignment engages students in the iterative design optimization process using previously developed parametric massing models. Using design exploration and optimization tools, students systematically evaluate various design possibilities. The assignment challenges students to define optimality amid a complex, often conflicting array of KPIs. Students iteratively revise model configurations and analysis workflows to meet their design objectives. Ultimately, they select one or more design schemes to develop further after discussing trade-offs in a studio review. Figure 19.7 illustrates the exploration and selection process for a project promoting active mobility at the Philadelphia waterfront. A Parallel Coordinate Plot is used to evaluate the relationships among various KPIs. It shows that certain KPI pairs, such as green space area and total visitor count, conflict because green spaces do not generate as much commercial activity and population density. Additionally, the relationship between other KPI pairs, such as total visitor count and high-footfall block frontage length, exhibits more complex dynamics due to the sensitivity of these KPIs to spatial factors like program configurations and network layouts.

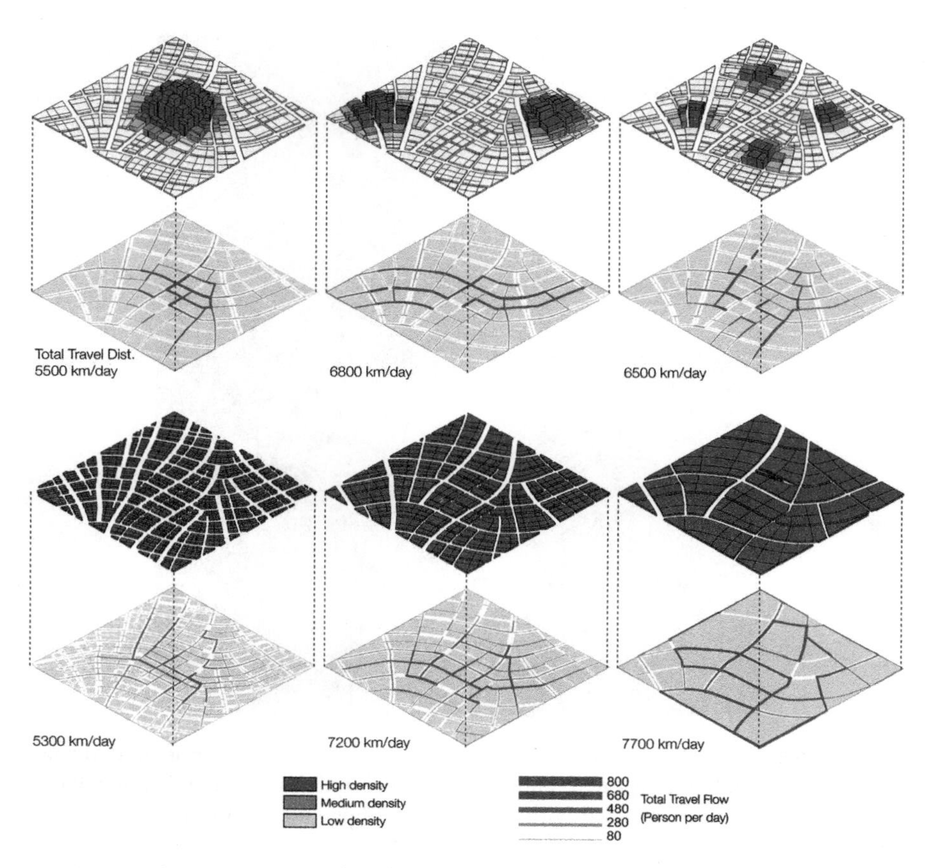

FIGURE 19.3 A parametric study illustrating the variations in total daily travel distance for local residents, resulting from different spatial configurations of urban densities and street network layouts. Each configuration shares the same total population, jobs, amenities and program types.

A4 Design Development and Delivery

The final assignment requires students to detail the design of buildings, streets and public spaces. Emphasis is on exploring effective methods of presenting design outcomes that convey a compelling, believable vision for the site. Students chose an area of focus, such as a building typology, a prototypical block with several buildings, or public spaces at the heart of their master plans. The development must demonstrate how the design advances the sustainability goals from previous assignments. In some cases, students also develop novel presentation techniques beyond conventional drawings and renderings, such as animations, interactive web interfaces and augmented reality, to offer a more comprehensive and immersive demonstration of the design, quality of space and spatiotemporal performance dynamics.

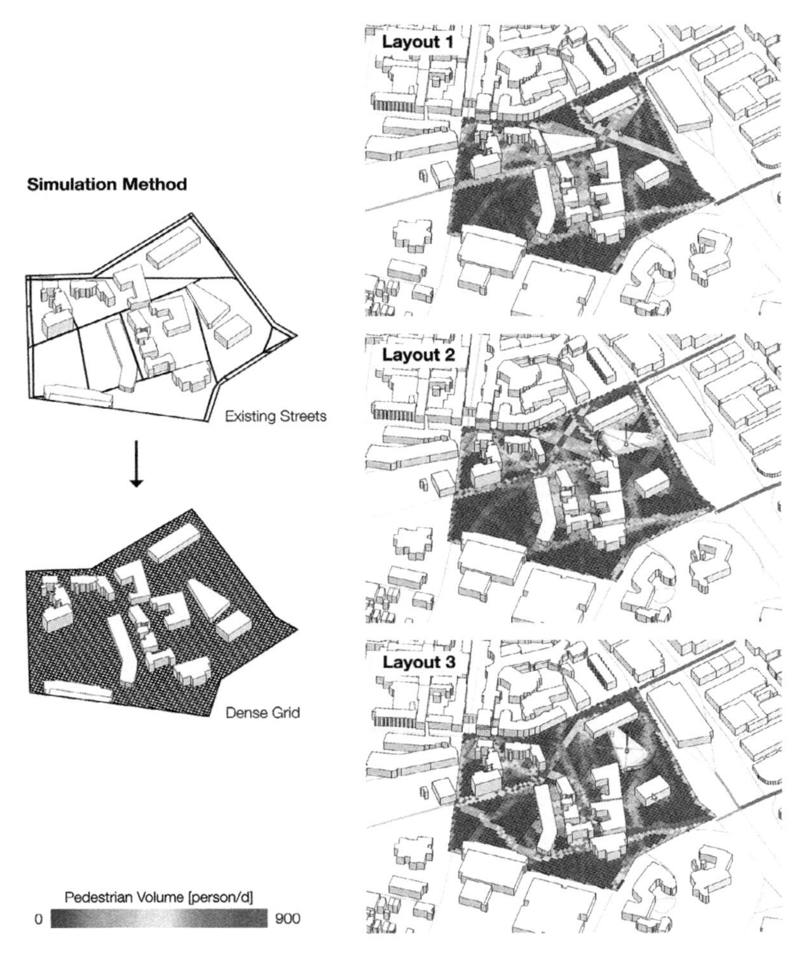

FIGURE 19.4 A novel approach utilizing the mobility simulation tool to visualize pedestrian flows in response to changes in building layout. The existing street network is replaced with a dense grid, enabling simulated pedestrians to move freely and allowing natural movement patterns to emerge. The behavioral assumption is made that pedestrians tend to choose the shortest path available.

Analytics

A1 Site Analysis and KPI Development

Urbano[8] serves as a pivotal tool for data-driven site analysis. It facilitates the straightforward search and download of large-scale geospatial data from open data providers like OpenStreetMap or municipal GIS portals and supports the import of user data in various formats. Its unique geometry-metadata data structure within the CAD environment enhances geospatial data synthesis, management and visualization.

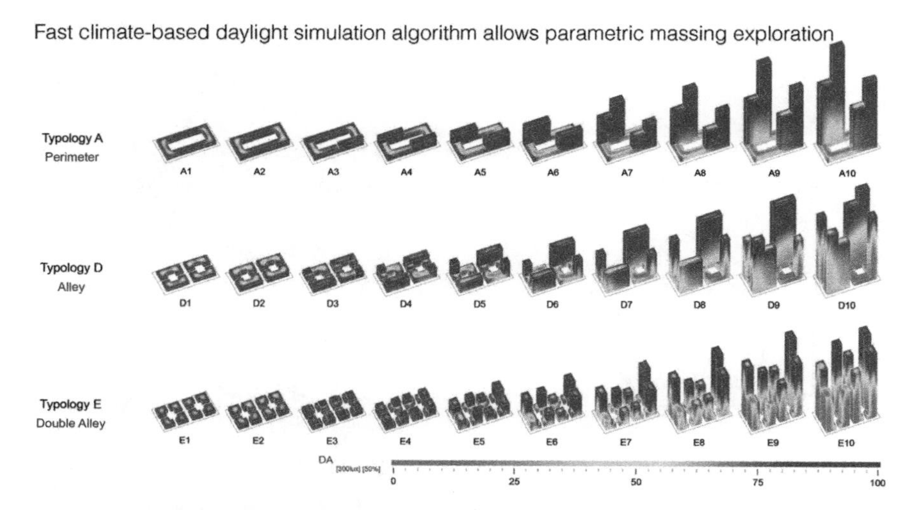

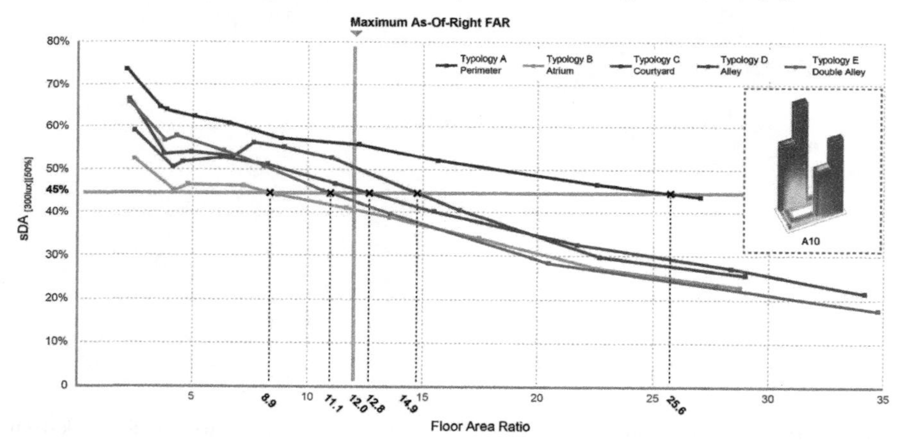

FIGURE 19.5 Variations in daylight accessibility and floor area ratio in response to changes in morphological parameters. Typology 10A significantly outperforms all other design families and allows designers to double the density before daylight performance drops below 45% sDA [300 lux, 50%].

Source: Saratsis E, Dogan T, Reinhart CF. Simulation-based daylighting analysis procedure for developing urban zoning rules. *Building Research & Information* 2017; 45(5): 478–491

A2 Parametric Investigation into KPIs

This assignment is underpinned by advanced simulation software that empowers students to compute environmental and mobility-related sustainability metrics. A work-in-progress version of Urbano and its Urban Building Energy Modeling (UBEM) extension is used for multi-modal mobility simulations and

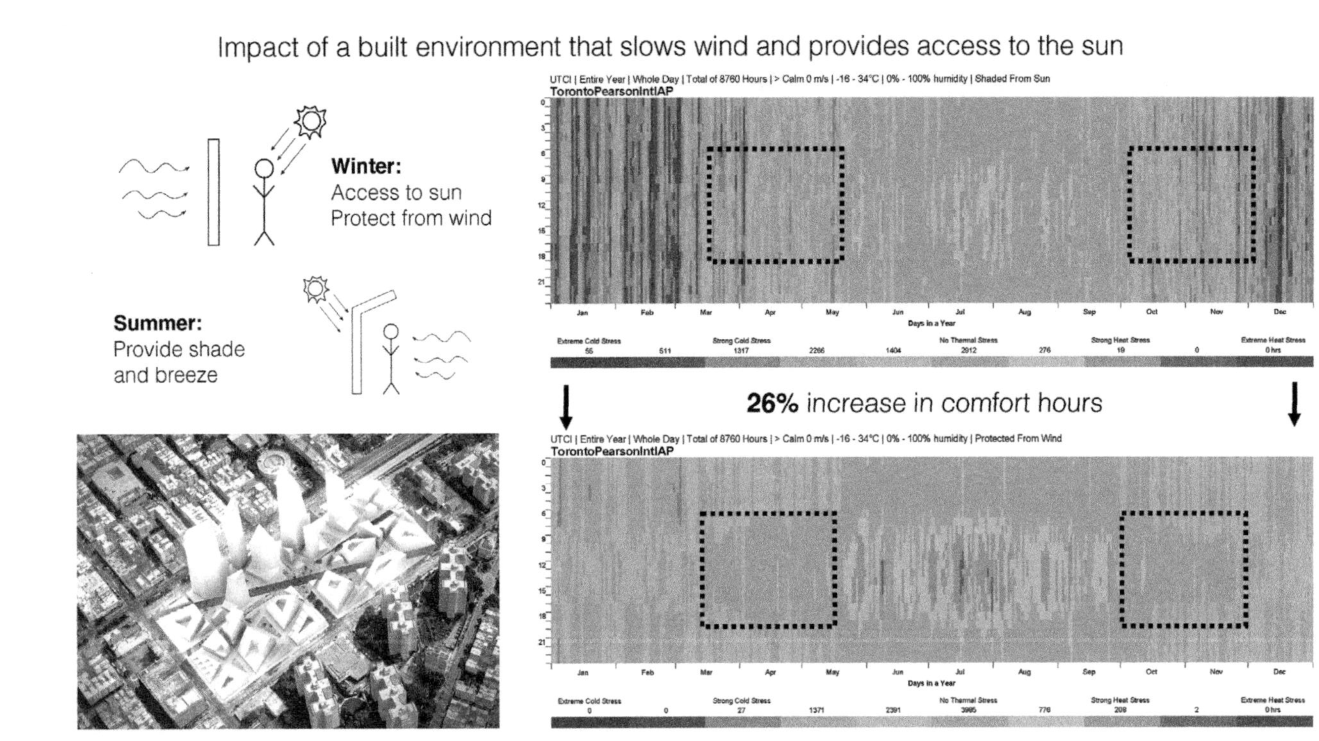

FIGURE 19.6 Outdoor thermal comfort as a design driver. Shaping of massing to allow for direct sunlight exposure on sidewalks and public spaces to enhance outdoor thermal comfort and to expand the thermally comfortable season by four months in the context of Toronto.

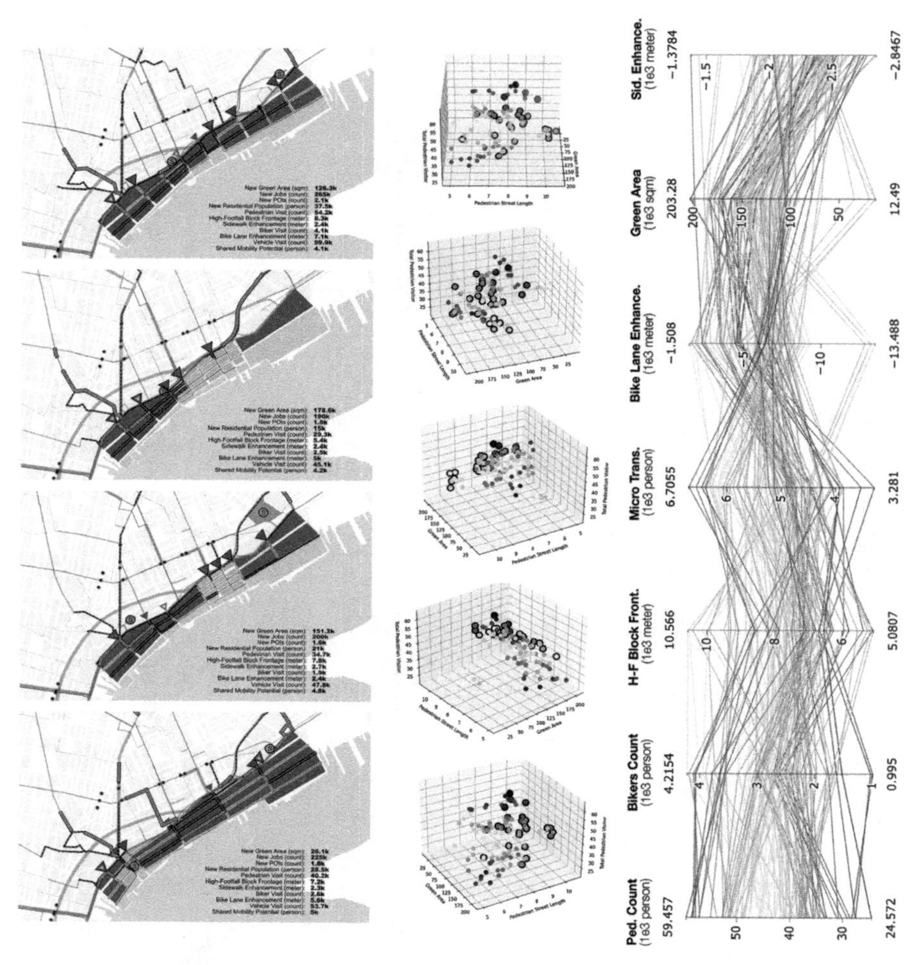

FIGURE 19.7 Design exploration of various master plan layouts at the Philadelphia waterfront site, utilizing multi-objective optimization techniques and tools, including Pareto Fronts and Parallel Coordinate Plots, to facilitate scheme comparison and optimality identification.

urban building energy analysis. Climate Studio (by Solemma LLC, http://solemma.com) is used to assess daylight availability within varying temporal and climatic contexts. For the evaluation of microclimate and thermal comfort, Eddy3D[9] offers a granular and efficient simulation environment to analyze the Universal Thermal Climate Index (UTCI), a composite metric that considers air temperature, wind, radiation and humidity. Furthermore, the studio develops generative workflows to create land use patterns, building forms and street networks.[10] These parametric design models are then passed and sometimes translated into the required simulation input format to compute the KPIs.[11] The students are encouraged to critique these established workflows and advance the studio's contributions to algorithmic thinking and computational innovation in urban design development

A3 Design Exploration and Optimization

In this phase, Wallace[12] is used for evolutionary design exploration and optimization, facilitating the iterative testing and refining of design solutions.

Required Readings

Hensen, J. L. M. and R. Lamberts (Eds.). *Building Performance Simulation for Design and Operation* (2nd ed.). Routledge, 2019. https://doi.org/10.1201/9780429402296

Yamagata, Y. and P. P. J. Yang. *Urban Systems Design: Creating Sustainable Smart Cities in the Internet of Things Era*. Elsevier, 2020.

Zhang, S. and L. Wilson. "Computational Urban Design." In *Machine Learning and the City*. John Wiley & Sons, 2022, pp. 393–407. doi: 10.1002/9781119815075.ch30.

Recommended Readings

Batty, M. *Inventing Future Cities*. MIT Press, 2018.

Haas, T., Ed. *Sustainable Urbanism and Beyond: Rethinking Cities for the Future*. Rizzoli, 2012.

Outreach Opportunities

The design studio facilitates various outreach opportunities that enrich student learning through direct engagement with the field of urban design and planning. To extend students' practical understanding of urban forms, functions, and their environmental and societal impacts, design practitioners are invited to participate in the education process through guest lectures, desk critiques, reviews and practitioner-led site visits to urban design projects. These activities have fostered collaborations and partnerships with the professional community, including leading firms like KPF, Sidewalk Labs, Gehl and municipal governments in Ithaca and Philadelphia, which offer students internships and other professional opportunities where they can leverage the unique skills they acquired in the course. The studio has also effectively translated students' research and design initiatives into academic contributions.[13] Furthermore, the studio has established connections with city planning departments and sustainability agencies, contributing to municipal government and policy to inform and align student project objectives with current urban planning challenges and sustainability goals. These connections often evolve into deeper research collaborations. For instance, the urban modeling and simulation techniques tested in this course underpin a research partnership with the City of Ithaca to create a digital twin platform to aid sustainable urban planning decisions. These collaborations foster mutually beneficial relationships between academic institutions and community organizations, wherein student involvement enhances human capacity and delivers essential skills.[14] Such partnerships help address staffing constraints in cities like Ithaca, which urgently require decarbonization initiatives but frequently lack the resources to plan these transformations using emerging computational technologies and data-driven processes.

Course Assessment

Student feedback from course evaluations is the primary assessment criterion and basis for iterative improvements to the studio. Students value the practical modeling and simulation skills acquired, noting their applicability to future projects and contribution to long-term professional growth. Furthermore, the extensive use of simulation tools and performance-based design approaches fosters innovation, enhances problem-solving skills and deepens their understanding of urban sustainability. Initially, students faced a steep learning curve in software workshops. Key technical challenges in urban-scale performance analysis in design included difficulty accessing data, defining specific criteria, employing or developing appropriate assessment tools, interpreting simulation results and integrating these results into master planning and architectural design. These challenges have driven the evolution of course tooling. A significant outcome has been the continuous development of Urbano, which aims to streamline urban data acquisition, synthesis and visualization within the CAD environment. By automating these processes, Urbano enables students to concentrate on innovative design and the rigor of their performance analysis rather than overcoming technical challenges. This significantly enhanced the learning experience.

This course touches key program criteria such as PC3 Ecological Literacy and Responsibility as well as PC5 Research and Innovation and fulfills various student learning objectives as outlined in NAAB accreditation. It equips students with an understanding of emerging technologies and the methods to rigorously evaluate the environmental performance of design proposals (SC.4 Technical Knowledge). Additionally, the course enhances students' capacity to make informed design decisions, considering user requirements, site conditions, measurable environmental impacts and quantifiable building performance outcomes (SC.5 Design Synthesis and SC.6 Building Integration).

Looking Ahead

Looking ahead, the anticipated innovations in visualization and dissemination can enhance how design outcomes and scenarios are engaged with and understood. For example, for visualization, the development of web interfaces for interactive scenario testing and data visualization is expected to increase the accessibility and usability of complex information significantly. Additionally, integrating virtual reality (VR) and augmented reality (AR) techniques could allow users to explore metrics and performance dynamics in real-time as design changes are made, offering a more immersive and intuitive understanding of spatial and environmental impacts. In terms of dissemination, there will be a move towards more inclusive and participatory processes. These will involve engaging a broader range of stakeholders and local communities through web interfaces, interactive workshops and targeted surveys. This approach could democratize the design process and ensure that the outputs are more comprehensively informed by diverse perspectives, ultimately leading to more sustainable and accepted urban solutions.

Notes

1 UNEP, *Sustainable Buildings and Climate Initiative*, www.unep.org/sbci/AboutSBCI/Background.asp (2015, accessed 5 August 2015).
2 IEA, *World Energy Statistics and Balances*, www.iea.org/data-and-statistics/data-product/world-energy-statistics-and-balances (2021, accessed 13 April 2024).
3 Dogan T, Kastner P, Tseng HM, et al. *Impact and Cost Analysis of Thermal Load Electrification Measures Using Automated Urban Building Energy Modeling in Ithaca, NY*. Denver, Colorado, 2024; Reinhart C, Dogan T, Jakubiec JA, et al. Umi – An Urban Simulation Environment for Building Energy Use, Daylighting and Walkability. In: *13th Conference of International Building Performance Simulation Association, Chambery, France*, 2013, pp. 476–483.
4 Dogan T, Reinhart C, Michalatos P. Urban Daylight Simulation: Calculating the Daylit Area of Urban Designs. In: *Proceedings of SimBuild*. Madison, Wisconsin, USA, 2012; Dogan T, Park YC. A Critical Review of Daylighting Metrics for Residential Architecture and a New Metric for Cold and Temperate Climates. *Lighting Research & Technology* 2018; 51(2) (doi:10.1177%2F1477153518755561); Dogan T, Park YC. Testing the Residential Daylight Score: Comparing Climate-Based Daylighting Metrics for 2444 Individual Dwelling Units in Temperate Climates. *Lighting Research & Technology* 2020; 52: 991–1008; Saratsis E, Dogan T, Reinhart CF. Simulation-Based Daylighting Analysis Procedure for Developing Urban Zoning Rules. *Building Research & Information* 2017; 45(5): 478–491.
5 Dogan T, Yang Y, Samaranayake S, et al. Urbano: A Tool to Promote Active Mobility Modeling and Amenity Analysis in Urban Design. *Technology|Architecture + Design* 2020; 4: 92–105; Yang Y, Samaranayake S, Dogan T, Assessing Impacts of the Built Environment on Mobility: A Joint Choice Model of Travel Mode and Duration. *Environment and Planning B: Urban Analytics and City Science* 2023. doi: 10.1177/23998083231154263.
6 Kastner P, Dogan T. Eddy3D: A Toolkit for Decoupled Outdoor Thermal Comfort Simulations in Urban Areas. *Building and Environment* 2022; 212: 108639.
7 Myrick Svante L. *Ithaca Green New Deal Summary*, www.cityofithaca.org/DocumentCenter/View/11054/IGND-Summary-02-11-2020 (2020, accessed 25 March 2020).
8 Dogan T, Yang Y, Samaranayake S, et al. Urbano: A Tool to Promote Active Mobility Modeling and Amenity Analysis in Urban Design. *Technology|Architecture + Design* 2020; 4: 92–105.
9 Kastner P, Dogan T. Eddy3D: A Toolkit for Decoupled Outdoor Thermal Comfort Simulations in Urban Areas. *Building and Environment* 2022; 212: 108639.
10 Sun Y, Dogan T. Generative Methods for Urban Design and Rapid Solution Space Exploration. *Environment and Planning B: Urban Analytics and City Science* 2022; 50(6).

11 Yang Y, Samaranayake S, Dogan T. An Adaptive Workflow to Generate Street Network and Amenity Allocation for Walkable Neighborhood Design, https://ecommons.cornell.edu/bitstream/handle/1813/69955/PublishedPaper.pdf?sequence=3.
12 Makki M, Showkatbakhsh M, Tabony A, et al. Evolutionary Algorithms for Generating Urban Morphology: Variations and Multiple Objectives. *International Journal of Architectural Computing* 2019; 17: 5–35.
13 Dogan T, Park YC. Testing the Residential Daylight Score: Comparing Climate-Based Daylighting Metrics for 2444 Individual Dwelling Units in Temperate Climates. *Lighting Research & Technology* 2020; 52: 991–1008; Saratsis E, Dogan T, Reinhart CF. Simulation-Based Daylighting Analysis Procedure for Developing Urban Zoning Rules. *Building Research & Information* 2017; 45(5): 478–491; Du P, Mavinkere NCR, Yang Y. Generative Urban Design: A Workflow Integrating Real-World Mobility, Zoning, and Multi-Objective Simulation. *ARCC 2023 Conference Proceedings* 2023: 413–420. https://www.arcc-arch.org/wp-content/uploads/2023/09/ARCC2023ProceedingsFINAL-PW.pdf.
14 Comeau DL, Palacios N, Talley C, et al. Community-Engaged Learning in Public Health: An Evaluation of Utilization and Value of Student Projects for Community Partners. *Pedagogy in Health Promotion* 2019; 5: 3–13.

20

DECARBONIZING CURRICULUM THROUGH ENVIRONMENTAL STEWARDSHIP AND EXPERIENTIAL LEARNING

Elizabeth Martin-Malikian

Course Overview

Course Name

Learn-by-Doing at Arcosanti: an urban experiment that practices sustainability through a live/work/learn micro-city

University / Location

The School of Architecture at Arcosanti / High Sonora Desert, AZ, USA

Targeted Student Level

Career beginners, pivoters, gap year students, non-traditional learners and environmental advocates and educators, who are interested in ecology, regenerative agriculture, design, architecture and urbanism. A background in the field is not required.

Course Learning Objectives

In this course, student workshoppers will:

- Understand the rationale behind arcology and the transformation of degraded landscapes into healthy, biodiverse habitats through ecological learning and hands-on stewardship in a built arcology.
- Learn about resource-conscious architecture terminology through readings, discussions and workshops that analyze carbon neutral design approaches.

DOI: 10.4324/9781032692562-26

- Develop and implement architectural and construction techniques centered on building a self-sufficient pedestrian city, including examination, demolition, remodeling of existing structures and the importance of creating functional community environments.
- Explore and apply the principles of sustainable food production, utilizing diverse growing techniques such as greenhouse agriculture, hydroponics, raised beds, orchards, field crops and permaculture food forests.
- Uncover, document and preserve the legacy of past work, including art, sketches, documents, models and photographs, understanding the importance of preserving and digitizing the evolution of emerging arcologies.

Course Learning Outcomes

Successful completion of a live/work/learn workshop at Arcosanti will enable students to:

- Develop an understanding of how visionary urban ideas have positively shaped professional practice, education and applied research by emphasizing the interconnectedness between natural + built environments + human behavior.
- Communicate key concepts of arcology and be able to apply them to environmental issues, as well as appreciate both the potential and limitations of the project.
- Obtain a deeper understanding of how human development impacts ecological footprint and systems.
- Understand the four central tenets: (1) embracing frugality & resourcefulness; (2) nurturing ecological accountability; (3) fostering experiential learning; and (4) leaving a limited footprint – that form the foundation of TCF's advocacy efforts (Figure 20.1) and has played a significant role in shaping the foundations work since day one.

Embracing Frugality & Resourcefulness

Thoughtfully approaching the built world and our practices in daily life in ways that are experientially rich but materially frugal.

Nurturing Ecological Accountability

Supporting the world's human habitat without burdening the Earth's capacity and limited natural resources.

Fostering Experiential Learning

Promoting the empowering impact of learning-by-doing and the dynamic educational experience that comes from immersion in one's work.

Leaving a Limited Footprint

Advocating for an urban density model where mixed-use spaces make better use of limited resources and connect to economic, social, and cultural outlets.

FIGURE 20.1 The Cosanti Foundation's four advocacy values that workshoppers will be exposed to during an immersive experience at Arcosanti.

Source: The Cosanti Foundation website, www.arcosanti.org

Course Methodology

Philosophy

> The discovery that the world is fragile is going to be probably the single most inspiring element for the new century for architecture. Not in terms of morality or consuming less energy but finding a new language that is actually the language of building that breathes ...
>
> – *Renzo Piano, architect*[1]

> [That this] off-grid "urban laboratory" would be self-sustained and collaborative at every level, from the production of food and building supplies to its construction, maintenance, and skill-share education module. It was nothing if not ambitious.
>
> – *Alice Bucknell,* Metropolis Magazine[2]

> There are ... aspects of Soleri's work I agree with and admire ... he started by attracting a group of people willing to work hard, a group held together by valuing ideas more than material rewards.
>
> – *Daniela Soleri, research scientist, UC Santa Barbara*[3]

> Our cities should be compact, cooperative, multifunctional, waste-minimizing environments that concentrate creative energies instead of disintegrating and degrading them in the energy sink of sprawl.
>
> – *Paolo Soleri, architect and urban planner*[4]

Since 1965, The Cosanti Foundation (TCF) has been dedicated to experiential learning as an approach to tackle ecological and sustainability challenges in the built environment. Although not a degree-granting institution, TCF has developed an extensive set of educational programs that help to increase a public understanding of creating balance between the built and natural environment (Figure 20.2). The foundation also collaborates with universities to offer customizable, hands-on workshops. As a pedagogy, participants (referred to as workshoppers or students) immerse themselves into the community of residents at Arcosanti, a live/work/learn prototype for an eco-friendly, micro-city that has been a work-in-progress for more than four decades. TCF emphasizes experiential learning, architectural immersion and a land stewardship program that grew out of the foundation's mission to demonstrate how urban conditions could be improved while minimizing the destructive impact on the Earth.[5] The project still serves as a global and educational training ground for architects, historians, environmentalists and ecologists who come to study arcology – the concept of *fusing architecture with ecology* – through workshops, research and programs. A workshop cohort brings together teachers, leaders and students passionate about inquiry-based learning based on four themes: environment and climate change; local materials

FIGURE 20.2 One of the original workshop posters 1963 (left), calendar promoting the Arcosanti workshops 1973 (middle) and recent workshop posters 2023 (right).

Source: The Cosanti Foundation – Archives at Arcosanti and education department

and silt casting; agriculture as a tool to achieve social sustainability; and the future of learning.

Definitions

Carbon neutrality – Arcology views carbon neutrality as integral to creating compact, self-sufficient designs that minimize environmental impact. Paolo Soleri coined the term in 1969 where architecture and ecology are two parts of the same thing, inseparable in their effect on humans and by extension Earth. Carbon neutrality and arcology intertwine in creating sustainable, self-sufficient architectural designs to minimize carbon emissions and pollution that is a function directly related to wastefulness. "The elimination of wastefulness is the elimination of pollution."[6]

Experiential learning – The process of learning by doing.

Miniaturization-complexity-duration (MCD) – A lean alternative principle that suggests the evolutionary engine of life over time is a means to understand societal systems through the efficiency and complexity of nature.

Nature-based solutions (NbS) – Actions that address broader societal challenges by protecting, sustainably managing and restoring ecosystems, benefiting both biodiversity and human well-being.[7]

Solver community – A diverse group of problem solvers collaborating to tackle challenges from fresh perspectives.

The urban effect – Based on nature where the organizing of cellular components grows into increasingly advanced forms of life resulting in the creation of an urban environment or cities. Building upon the MCD principle, it carries a set of implicit values through a density of living, a density of habitat and the results that flow from its interaction.[8]

Implementation

General

There are a variety of different workshop formats from a 1-day to 6-week program as well as customizable multi-day options in collaboration with universities and community partners. In single day or weekend programs, participants learn a variety of practices and crafts through hands-on experience (Figure 20.3). In the multi-week programs, workshoppers focus on one of the four primary experiential learning opportunities: (1) field and work-based; (2) project and problem-based; (3) community and public outreach-based; and (4) archival and research-based.

Arcosanti itself is a demonstration project that student workshoppers use as a 1:1 full scale, living, breathing precedent to engage and learn. Typically, a workshop starts with a week-long, structured "seminar week" introducing participants to the history of the project with a presentation of the ideas and construction of Arcosanti along with a tour of the archives, a treasure trove of

FIGURE 20.3 Children from ages 8–14 learning about sustainability and hands-on craft at the Cosanti by Design workshop.

Source: Photographs by the author and TCF education department

ideas, drawings and models considered the heart of the project. Primarily working outside, participants are familiarized with all the departments at Arcosanti including hands-on demonstrations of welding, along with slit-casting techniques used in ceramics, sand-casting used in the bronze foundry, a tour of the agriculture beds, fields and food forest, as well as go on an ecological nature hike along the Agua Fria River and learn about riparian habitats.

Experiential Learning Workshop Approach

Described by *New York Times* architecture critic Ada Louise Huxtable as an "urban laboratory,"[9] Arcosanti is an experimental micro-city focusing on the importance of environmental accountability (Figure 20.4). The project aims to showcase how cities can coexist with the natural landscape while engaging the community in living, working and learning together. Through hands-on experiences and reflection, workshop participants gain valuable skills and connect

FIGURE 20.4 The goal of the Arcosanti community is to explore the concept of "arcology," which combines architecture and ecology. The concept was conceived by Italian-American architect Paolo Soleri, who began construction in 1970 as part of a workshop program to demonstrate how urban conditions could be improved while minimizing the destructive impact on the earth.

Source: Carol M. Highsmith from American Library of Congress Collection 2018

conceptual ideas to real-world situations. The workshop tasks and discussions are organized into five tactics aimed at creating a more resilient future, starting with an introductory seminar week and continuing throughout the immersive experience: (1) Arcology Intention and Practice; (2) Approach to Sustainable Urban Development; (3) The Urban Effect; (4) Hyper-structure and Lean Alternative; and (5) Solver Community.

Arcology Intention and Practice

- What alternative approaches or solutions does arcology consider regarding consumption of resources and unsustainable expansion?
- How can the intention of arcology be practically implemented at a city level?
- What specific policies and practices contribute to unsustainable urban sprawl?

The concept of arcology underscores the urgent need to achieve harmony between architecture and the natural world. It calls for a shift in perspective, moving beyond individual buildings to consider not only reducing consumption but also responsibly managing the land and conserving natural resources. In stark contrast, urban sprawl's relentless expansion, vast distances between locations, and excessive energy and resource use have caused significant harm to the environment and society, while also proving to be impractical. This unsustainable pattern has resulted in severe resource waste and environmental damage (Figure 20.5). Despite the obvious negative effects, short-sighted policies and practices continue to drive urban development in a direction that is detrimental to addressing climate change. To achieve carbon neutrality within the next 25 years, students are interested in exploring how the principles of arcology can help communities adapt to the already entrenched socio-ecological impacts of global warming. Arcology, as a concept, offers a comprehensive framework primed to sustain and inspire[10] our efforts as we strive to enhance health and well-being.

Approach to Sustainable Urban Development

- What innovative ideas does arcology bring to urban development, and how does it differ from traditional approaches?
- In what ways does Arcosanti embody the principles of arcology?
- How can Arcosanti serve as an educational resource for understanding urban development?
- In what ways can modern cities minimize their land use while maximizing their positive environmental impact?

There is a pressing need to decarbonize the built environment making the goal of transforming the urban development paradigm a challenging, but worthwhile goal. Arcology as an urban approach offers a hopeful vision proposing that cities be vertical, dense, optimized for pedestrians, and in balance with their natural

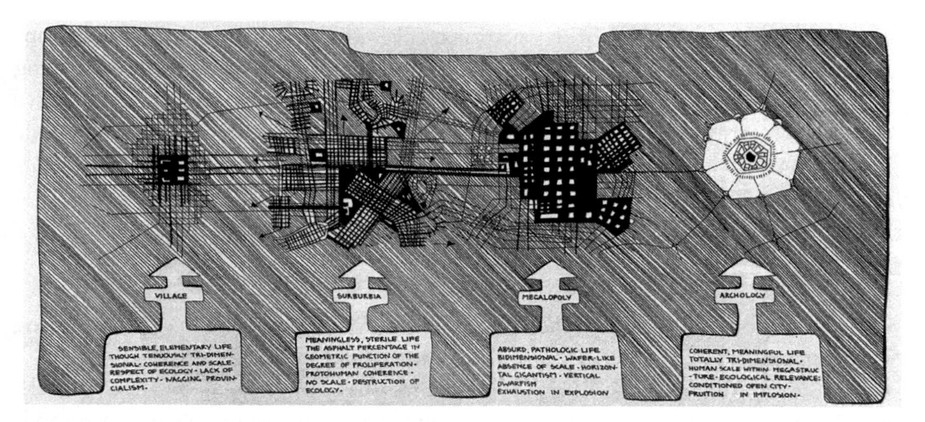

FIGURE 20.5 Diagram illustrating urban development considering "The Automobile Mystique and the Asphalt Nightmare" from *Arcology: City in the Image of Man*, MIT, 1969, p. 23.
Source: The Cosanti Foundation – Archives at Arcosanti

surroundings. While studying at Taliesin, Paolo Soleri challenged Frank Lloyd Wright's auto-centric approach to US city planning. In response, he created Arcosanti as a prototype for a city that embodies self-sufficient, low-impact living that is in harmony with the natural landscape and offers a contrast to sprawling urban-suburban development.[11] During workshops (Figure 20.6), participants actively engage in on-site learning to envision a city from three perspectives: the position of nature (objective), the position of man (subjective), and the position of neonature (the man-made)[12] This experiential approach makes Arcosanti a learning tool for understanding how to: (1) visualize and experience a 1:1 scaled, physical model of the city of the future; (2) grasp the effects of sprawl and why a new infrastructure model is needed to reduce distance between activities; and (3) recognize critical development and transition phases to achieve a healthier ecological outlook.[13]

The Urban Effect

- How can spaces be effectively designed to promote the urban effect?
- What are some specific examples of spaces that can foster the urban effect in cities?
- How does the concept of minimizing distance and duration contribute to creating vibrant and dynamic urban communities?

In cities around the world, we can see the positive outcomes of combining different functions and activities within urban areas. These mixed-use spaces become shared, public areas where social interaction and growth thrive. According to an arcological approach, for a city to be successful, it needs to be less dependent on the automobile, have a minimal ecological impact and

FIGURE 20.6 Soleri's arcology concept is being put to the test in the Arcosanti experimental community in Arizona after the silt-casting technique was perfected at Cosanti. The process was well-suited to a workshop environment, which included a combination of skilled and unskilled participants. Buildings were composed of locally produced concrete and designed to capture sunlight and heat.

Source: Arcosanti 1970, Ivan Pintar (top left), Cosanti 1964, William E. Tickel (top right), and Arcosanti ext. 1970s from The Cosanti Foundation – Archives at Arcosanti (bottom)

support life processes. When these areas are in harmony with nature, the city can be more than just the sum of its parts. Soleri argues that the principle of miniaturization-complexity-duration (MDC) is not simply about creating small structures in close proximity but about maximizing interaction and exchange, while decreasing travel distance between activities. MDC relationship principle argues that the advent of cities was not the origin of urban life; instead, it was the urban effect, the organizing of subcellular components and cells into increasingly advanced forms of life, that ultimately impelled the founding of cities.[14] In an organic compact structure, if it doesn't promote or enable the urban effect, it simply becomes a physical structure. Although miniaturization is an important concept in arcology, it is meant to be a design tool to create "a lasting solution for land conservation,"[15] with the ecological intent to confront sprawl and activate a more compact urban reformation.

Hyper-structures and Lean Alternative

- What are the implications and benefits of hyper-structures for urban living and environmental sustainability?

- What are the potential challenges of implementing hyper-structures at such a large scale?
- What are the design considerations involved in creating lean pedestrian-oriented communities prioritizing mobility, access and resource conservation?

The concept of the lean alternative offers a transformative solution to counter the current unsustainable development paradigm. By strategically constructing dense, self-contained hyper-structures along major transit routes and directing development towards them, we can create space for nature to thrive in the surrounding areas.[16] These cities functioning as hyper-structures mimic nature by building on or over existing structures without disrupting them, maintaining a delicate balance between technology and nature, and allowing natural processes to gradually transform the environment. The key advantage of the lean alternative lies in the creation of strategically located habitats that will help reduce our heavy reliance on fossil fuels. This shift embraces frugal resiliency, environmental sustainability and ecological balance, marking a departure from a hyper-consumptive lifestyle to one emphasizing conscious resource utilization, akin to natural operations.

Solver Community

- In what ways do teams tend to outperform individuals in generating solutions?
- What essential interpersonal skills are developed through a team-based approach?
- Can a solver community transform and improve design education?
- Why is bringing diverse perspectives to a problem essential for understanding its full complexity?

When envisioning the future of learning, Arcosanti stands out for its innovative approach in establishing a collaborative community that comes together to tackle and resolve challenges. The innovative solver community approach shows great potential for the learn-by-doing workshops at Arcosanti, as it challenges the traditional passive learning model. Workshoppers are tasked with addressing meaningful problems and navigating community issues, instead of completing standard assignments, fostering a project-driven way of learning. Moreover, a solver community necessitates collaboration, problem debate, criteria setting for assessing different perspectives and the development of detailed solutions.[17] When executed effectively, immersive environments have the potential to greatly enhance the overall learning experience for participants. The workshop experience being immersed within a community of residents naturally creates a real-life scenario: an endeavor to create a different urban prototype guided by the foundation's dedication to social and environmental equity and justice.[18] Through this solver community approach, workshop participants can systematically consider challenges, identify underlying problems, evaluate alternative approaches and generate solutions.

Required Reading

Grierson, David. "Unfinished Business at the Urban Laboratory – Paolo Soleri, Arcology, and Arcosanti." *Open House International*, vol. 41, no. 4, 1 Dec. 2016, pp. 63–72, https://doi.org/10.1108/ohi-04-2016-b0009

McCullough, Lissa, Ed. *Conversations with Paolo Soleri*. NY: Princeton Architectural Press, 2012.

Soleri, Paolo. *Arcology: The City in the Image of Man*. Cambridge, MA: MIT Press, 1969.

Soleri, Paolo. *The Sketchbooks of Paolo Soleri*. Cambridge, MA: MIT Press, 1971.

Soleri, Paolo and Scott M. Davis. *Paolo Soleri's Earth Casting*. Salt Lake City, UT: Peregrine Smith Publishers, 1984.

Recommended Reading

Agrest, Diana, et al. *Architecture of Nature: Nature of Architecture*. Novato, CA: Applied Research and Design Publishing, 2018.

Balsas, C. J. L. (2020). "Paolo Soleri and America's Third Utopia: The Sustainable City-Region." *Journal of Urbanism: International Research on Placemaking and Urban Sustainability*, vol. 13, no. 4, pp. 410–430, https://doi.org/10.1080/17549175.2020.1726798.

Beatley, Timothy. Biophilic Cities: *Integrating Nature into Urban Design and Planning*. Washington, DC: Island Press, 2011.

Friedman, Thomas L. *Hot, Flat, and Crowded: Why the World Needs a Green Revolution, and How We Can Renew Our Global Future*. London, Penguin Books, 2009.

Helm, Dieter. *Natural Capital: Valuing Our Planet*. New Haven, CT: Yale University Press, 2016.

Klein, Naomi. *This Changes Everything: Capitalism vs. the Climate*. New York, Simon & Schuster, 2014.

Luke, Tim. "The Politics of Arcological Utopia: Soleri on Ecology, Architecture and Society." *Telos*, vol. 1994, no. 101, 1 Oct. 1994, pp. 55–78, https://doi.org/10.3817/0994101055.

Munro, Karen, and David Grierson. "Towards the Development of a Space/Nature Syntax at Arcosanti." *Open House International*, vol. 41, no. 4, 1 Dec. 2016, pp. 48–55, https://doi.org/10.1108/ohi-04-2016-b0007.

Tuana, Nancy. *Racial Climates, Ecological Indifference*. Oxford: Oxford University Press, 2023.

Walker, B. H., and David Salt. *Resilience Thinking: Sustaining Ecosystems and People in a Changing World*. Washington, DC: Island Press, 2006.

Young, Gabriel Yit Viu. "Climate-Resilient Urban Design from a Biomimicry-Arcology Perspective." *International Journal of Environment, Architecture, and Societies*, vol. 2, no. 1, 28 Feb. 2022, pp. 1–15, https://doi.org/10.26418/ijeas.2022.2.01.1-15.

Outreach Opportunities

Beginning with a design/build studio at Arizona State University in 1963, TCF has established numerous meaningful outreach opportunities with professional groups, educational networks and independent scholars at Arcosanti and Cosanti. The intention is to continue to foster mutual learning and engage in meaningful dialogue in the context of climate innovation, land

FIGURE 20.7 ASU department of textiles and ceramics at Cosanti and Arcosanti (top). As an industry partner, Arcosanti is one of three sites that Michael Kotutwa Johnson will be using to grow Hopi corn and beans to revitalize Indigenous food ways – an effort supported by a grant from The Rockefeller Foundation's Climate Exploration Fund (bottom).
Source: Photographs by the author and TCF education department

stewardship, and the nature of making and meaning (Figure 20.7). The resulting community-centered opportunities, industry-partner projects and corporate-research initiatives open avenues to provide ideas, staff and resources to address the environmental challenges we are facing today. The foundation continues to engage in conversations about environmental issues through exhibitions, professional conferences and colloquiums, such as the OpenCitizen Gathering, Building Technology Educators Society (BTES), PACT: Transit + Community, HATCH/Catalyst 2030, to mention a few.

Course Assessment

Over the past sixty years, the workshop program has undergone various transformations, always with the shared goal of contributing to the community through use, program, value, or knowledge construction. Following an introductory seminar week, participants engage in experiential learning across four main categories: (1) field and work-based; (2) project and problem-based; (3) community or public outreach-based; and (4) archival and research-based, depending on individual goals and interests (Figure 20.8). Based on this unique pedagogy, workshop assessment is closely tied to reflecting on the contributions made to project advancement, increased skill building and personal discovery. Feedback is gathered through exit meetings and surveys to gain insight into how participants connect their personal interests with practical skills and knowledge building. The workshop concludes with a public presentation and celebratory gathering, marking the participants' transition to becoming an alumnus of the program.

FIGURE 20.8 Students participate in a variety of learn-by-doing activities.
Source: Photographs by the author and TCF education department

Lessoned Learned

Upon completion of the program, alumni carry their newfound knowledge into their professional endeavors, while others have the unique opportunity to exchange volunteer work for housing or join the Foundation's staff. This helps create a dynamic environment where people live, work and learn together, driving the project forward. Throughout the experiential learning process, participants are actively posing questions, investigating, being curious, assuming responsibility, being creative and gaining a deep appreciation of community and a broader view of the world. They also gain valuable insight into their personal skills, interests, passions and values.

Looking Ahead

Arcology is a realistic concept that initially faced strong resistance despite its logical environmental argument to help reverse the impact on climate. Today with all the environmental challenges the world is facing, universities interested in working towards creating a healthier planet can collaborate with the Foundation for single or multi-day experiential learning and creative problem-solving activities. Projects to examine arcology through the lens of other nature-based solutions like biomimicry, restorative agroforestry and Indigenous farming practices can also help provide students with valuable skills and tools to help understand how to protect habitat loss and slow down climate change. Carson Chan, Director of the Emilio Ambasz Institute for the Joint Study of the Built and the Natural Environment at MoMA, describes environmental architecture as an emerging field, where the contours of its history are just beginning to take shape.[19] By stepping outside traditional classrooms, students and faculty can together learn side-by-side with residents, and staff can show how past ideas of sustainability can inspire these "emerging arcologies" and work towards our common global goals of being carbon neutral by 2030 and fossil fuel-free by 2050.

For more information, email workshop@arcosanti.org.

Notes

1 Renzo Piano interview by Charlie Rose, June 5, 2008.
2 Alice Bucknell. "Complicated and Contradictory Legacy of Arcosanti." *Metropolis Magazine*, December 22, 2017. https://metropolismag.com/viewpoints/complicated-legacy-arcosanti-paolo-soleri/
3 Daniela Soleri. "It's You, Him, and His Work." *Medium*, February 28, 2017. https://medium.com/@soleri/sexual-abuse-its-you-him-and-his-work-88ecb8e99648
4 Paolo Soleri. *Arcology: The City in the Image of Man*. Cambridge, MA: MIT Press, 1969.
5 The Cosanti Foundation website. www.arcosanti.org
6 Paolo Soleri. *Arcology: The City in the Image of Man*. Cambridge, MA: MIT Press, 1969, p.39.
7 The Nature Conservancy website. https://www.nature.org/en-us/what-we-do/our-insights/perspectives/ncs-principles/
8 Lissa Mccullough (ed.). *Conversations with Paolo Soleri*. New York: Princeton Architectural Press, Cop, 2012. https://www.organism.earth/library/document/conversations-with-paolo-soleri
9 Ada Louise Huxtable. "Architecture Prophet in the Desert." *NYT*, March 15, 1970. https://timesmachine.nytimes.com/timesmachine/1970/03/15/93880110.html?pageNumber=118
10 Paolo Soleri. *Arcology: The City in the Image of Man*. Cambridge, MA: MIT Press, 1969, p.35.
11 *The Ideology of Paolo Soleri*, Arcosanti website. https://www.arcosanti.org/arcology/

12 Paolo Soleri. *The Sketchbooks of Paolo Soleri*. Cambridge, MA: MIT Press, 1971, p.4.
13 Paolo Soleri. *Arcology: The City in the Image of Man*. Cambridge, MA: MIT Press, 1969, pp.39–40.
14 Lissa Mccullough (ed.). *Conversations with Paolo Soleri*. New York: Princeton Architectural Press, Cop, 2012. https://www.organism.earth/library/document/conversations-with-paolo-soleri
15 Paolo Soleri. *Arcology: The City in the Image of Man*. Cambridge, MA: MIT Press, 1969, p.39.
16 Paolo Soleri, Youngsoo Kim, Charles Anderson, Adam Nordfors, Scott Riley, and Tomiaki Tamura. *Lean Linear City: Arterial Arcology*. Edited by Lissa Mccullough. Mayer, AZ: Cosanti Press, 2012.
17 Steven Mintz. "Seize the Power of Experiential Learning." www.insidehighered.com, July 26, 2021. https://www.insidehighered.com/blogs/higher-ed-gamma/seize-power-experiential-learning?v2Steve
18 Cosanti Foundation website. www.arcosanti.org
19 Carson Chan and Matthew Wagstaffe. *Emerging Ecologies: Architecture and the Rise of Environmentalism*. New York: MoMA, 2023, p.15.

CONCLUSION

Robin Z. Puttock

Pedagogy is personal. Watch ten design studio professors teach and you will, most likely, see ten unique design studio methodologies.

As stated in the Introduction, this book was conceived to highlight the many existing diverse, innovative and award-winning carbon neutral design methodologies currently implemented at architecture schools across North America. As a professor with experience at several architecture schools, and as a volunteer actively engaged on a national and international level, I saw a glaring need for increased accessibility to this course content, for students, of course, but also for educators interested in creating courses or adapting existing ones. A significant strategy to help combat the climate crisis is to educate as many current architecture students as possible about the role of the built environment in carbon emissions. This can be achieved if more professors are encouraged and empowered to do so. This compilation of pedagogies is designed to support this part of the solution.

When I began to consider prospective authors for this volume, I thought about what we, as educators, do. We inform and inspire. Therefore, in order for the book to best inform, I sought out leaders in carbon emission reduction pedagogy. Acknowledging that to be only half of the equation, I also actively sought out diverse perspectives in order to connect pedagogically with as many educators and administrators as possible. Like I said above, pedagogy is personal. The goal was to offer a buffet of strategies and tactics which could be adopted and adapted. The more creative, the better. Provocative? Yes, please! Controversial? Bring it on!

Ironically, in my quest for unique pedagogical strategies and tactics, while reviewing the chapter drafts, it quickly became evident that there are many commonalities that also unite this diverse group of thinkers. We agree that courses should start with the moral imperative. From a student's perspective,

DOI: 10.4324/9781032692562-27

this might mean: "Why are we learning this?" or "What impact can we have?" Another common theme is that the next phase of course content following the ethical argument, the courses typically include some sort of pre-design analysis. This can mean precedent, program and site analysis. It can also mean taking the time to understand your client and the specific problem and/or design challenge. Most, if not all, of the chapters describe the soft skill of developing relationships: relationships with clients, with professors, with fellow classmates. Group work and collaboration are critical as they reflect conditions found in the "real world". And finally, most chapters describe the importance of allowing all of the above to inform the overall goal-setting of the course. Themes of motivation, knowledge and human relationships. In other words, inform and inspire.

The realization that there were many commonalities among the diverse group of educators was energizing. In a way, I felt validated in my own pedagogy, as I hold each of these contributors in such high regard. However, at the same time, another irony materialized. Even though many commonalities emerged, my initial goal of highlighting and celebrating diverse viewpoints, which is clearly evident in these pages, also forced me reconsider many of my own views. I found myself reflecting on the big picture, considering ways to adapt my methods accordingly, and, in some cases, even outright abandoning them in favor of new ones. This was unexpected. My goal was to compile this volume to share the pedagogies with others, those who are perhaps less familiar with decarbonization course content or methodologies. I was not initially thinking of the benefit it could have for me, or others like me, who have been teaching this content for years. It is now a new hope of mine, after reading the final versions of each chapter, that educators of all levels of comfort teaching decarbonization content will read this book and be inspired to reflect upon or consider adapting their existing pedagogy, or perhaps even adopting new methods described in this book.

Having now reflected on the diverse contents of this book and its unexpected common themes, it is imperative to now consider where we go from here. Perhaps my two most significant insights are:

1. the need for organized advocacy for NAAB to include specific energy and carbon language in its accreditation criteria; and
2. the need for additional building reuse pedagogical resources.

As mentioned numerous times previously, this volume's intent was to be a buffet, not a deep dive into any singular topic; however, it is my hope that these two topics gain traction both in terms of pedagogical resources and advocacy opportunities.

Speaking of where we go from here, I will address the elephant in the book: AI. We are clearly at the beginning of a significant paradigm shift in education. How will it transform academia? It is too soon to tell. I only started to incorporate graphic and written AI software into my teaching and research in the fall of 2023. So far, it has been a powerful and exciting tool for both my students and me. In fact, in an effort to ensure I have a contemporary, if not ephemeral, conclusion for this book about innovative, provocative and sometimes controversial pedagogies, I am using it to write the closing thoughts of this conclusion, which seems, to me, appropriate:

> This book emphasizes the urgent integration of sustainable practices into architectural education, focusing on decarbonization. The chapters highlight the importance of hands-on, community-engaged learning to empower both students and local communities with practical sustainable solutions. Technological advancements such as computational tools and AI are seen as critical in advancing architectural design and performance analysis. Collaboration across disciplines is advocated to develop holistic solutions that integrate diverse perspectives and expertise. Overall, this is a clear evolution in pedagogical approaches within architectural education, aiming to prepare students for addressing global challenges through innovation and sustainable design principles.
>
> – *ChatGPT's response to "In five sentences, please summarize all 20 of the 'Looking Ahead' sections in this book"*

EPILOGUE

Legacy – A Dialogue among the Faculty of Clemson's Environmental Justice and Design Studio

Ulrike Heine, David Franco and George Schafer

Note from the Editor: *As I described in the Preface and Introduction to this volume, the primary goal for this heterogeneous compilation of architecture design studio methodologies is to encourage wide accessibility to and acceptance by faculty and administrators developing architecture curricula in an effort to maximize the impact on curtailing carbon emissions resulting from the construction and operations of our built environment. I had initially invited Ulrike Heine, David Franco and George Schafer, a power-house faculty team from Clemson University, to contribute a chapter to this effort, describing their unique pedagogy which I had followed and admired for years. However, while in discussions about the chapter framework, it became clear that perhaps a different approach to sharing their methodology, one that celebrates the many legacies of this pivotal studio, might be more appropriate. Thus, the idea for this epilogue was born. I am pleased to introduce this unique dialogue and I am grateful to the Clemson team for sharing it with all of us.*

Course Overview

Course Name

ARCH 8510: Design Studio III

Targeted Student Level

- Graduate
    - 1st year, 1st semester, Master of Architecture, 2-year program
    - 2nd year, 1st semester, Master of Architecture, 3-year program

DOI: 10.4324/9781032692562-28

Dialogue

From 2015 to 2022, a tight team of faculty at the Clemson School of Architecture co-taught a graduate studio organized around the AIA COTE Top Ten for Students competition. Over these seven years, this collaborative effort resulted in 22 awards, including 11 AIA COTE Top Ten for Students awards, a regional exhibition, and a video installation in Venice in the summer of 2023. More significantly, it has transformed how we at Clemson teach and learn about architecture and the environment. Our studio became an enjoyable place of intellectual exchange and dialogue among students and faculty, open to discussing social, environmental and political issues. In January 2024, the core team of the studio – Ulrike Heine, David Franco and George Schafer, who have since moved on to other roles and will no longer be co-teaching this course – sat together for a last time to reflect on the history of the course and the aspects that led to it being considered an exemplar of design for environmental justice instruction.

GEORGE SCHAFER: Ulrike and David – you originated this course, and both taught for the first two years before I joined you in year three. Perhaps we should begin with the origins of the course and its position within the broader graduate program at Clemson.

ULRIKE HEINE: I think there are three topics to touch on here: the curriculum, the AIA COTE Top Ten for Students competition and the overarching field. Clemson's Fluid Campus, where graduate students study abroad during the penultimate semester, requires us to put a lot of coursework in the bookended semesters of our curriculum, so students can travel during their chosen semester. The intention was to create a studio that sets the tone for how and what we teach at Clemson and what our expectations are from the students. We chose to implement environmentally responsive design early in the curriculum to raise the awareness of today's environmental challenges and our responsibility as architects to help solve these problems. And honestly – as an evaluation tool – nothing works better than exposing the course outcomes in the context of a national competition to assess the efficacy of the teaching concept.

GEORGE SCHAFER: When the course was originally launched as a tone-setting studio that highlighted issues of environmentally responsive architecture within a national competition, was it conceived of as a stand-alone studio or one that was one element within a suite of courses that helped scaffold the themes embedded in the AIA COTE Top Ten for Students competition?

ULRIKE HEINE: In the beginning, it was a stand-alone studio. And then we realized that the studio work is getting even stronger if it would be served by other courses, such "Research Methods" and "Production and Assemblies" courses. We learned that when students are focused on their studio projects in these supportive courses, it resulted in projects that were more conceptually and technically resolved.

DAVID FRANCO: Within the main structure that Ulrike pointed out, I would say that engaging with a competition in the studio was critical in the beginning because we understood that if it included a technical component, such as the AIA COTE Top Ten for Students competition did, it would provide a centrality to design – as a moment, as a place where all the other areas regarding architecture meet and overlap. For us, that was very important. Due to our backgrounds, both of us coming from the Polytechnic European schools, there's no strict separation between design and other fields of the built environment, from the urban to construction to wall section details. And at the same time, it provided an important connection to what the profession is actually about. It created this multidirectional connectivity between the technical disciplines and the reality of the profession. And then, too, the ambition – the conceptual ambition that a competition requires.

GEORGE SCHAFER: Maybe we could talk a little bit more about how setting it in the context of a competition also added pedagogical value to the studio. For example, the way that the competition imbued rigor into the course, which I think is undeniably true. It provided a framework for course longevity, where new students, critics and instructors join the course and bring their own perspectives into that conversation. There's a structure and stability inherent to the AIA COTE Top Ten for Students competition that allows for a lot of fluidity as the course advances. And there's a really important component of student motivation attached to being embedded in a competition, particularly since the course at Clemson has netted so many successful entries over the years.

ULRIKE HEINE: As we started showing that we could actually win in the AIA COTE Top Ten for Students competition there is the aspect of motivation for students being involved in that process over time. I think there's also the aspect of how to express yourself graphically without having the chance to present or explain it verbally. But knowing how to tell the story of your project to a point that someone, who has never seen it before, will understand it immediately, with not a lot of real estate for telling that story – that is a beautiful challenge. And keep in mind we're talking about Master of Architecture students from both the 2-year and the 3-year program. Sometimes half of the students in the studio don't have an architectural background. So, they have to find a way to express themselves in their drawings to the point that there is no level of misinterpretation.

DAVID FRANCO: Also, the question of the competition almost forces you into a point of view of innovation and challenging the status quo, what is normal, what are conventional assumptions. And, in this case, a competition about environmental design, of architecture that deals with the environment. It puts us in a place where we could react to the dissatisfaction with the typical solutions for so-called sustainable architecture: Active systems and simply, the addition of technological elements that are not connected to the form. The deep dissatisfaction with standard environmental architecture forces us to look at something we don't know and ask ourselves what it can be. And that gives the competition a really powerful role through the idea that we can motivate the students to invent environmentally responsive architecture. You don't need to fulfill the minimums or the technical requirements. You can reinvent an architecture that deals with the environment.

GEORGE SCHAFER: I agree. One of the strengths of how this course positions the measures embedded within the AIA COTE Top Ten for Students competition, and more broadly the AIA Framework for Design Excellence, is within a laboratory for testing ideas and hypotheses. Another important aspect about the competition is that it broadens the impact of the studio. With each successful student entry, the design ideas developed in the studio are being seen by a wide audience. And I think that's critically important for a forward-thinking, research-based studio that wants to extend the conversation beyond what environmentally responsive architecture means now.

ULRIKE HEINE: In fact, if the students want to win, they must create really good architecture. It's about exceptional concepts, exceptional execution. It's not about creating an average design by simply adding active systems and testing through software as a recipe for a very efficient building. Being in a competition necessitates outstanding design. And that is the essence of our studio. We're not sitting around strategizing who's on the jury and how can we notch another win. The competition just provides the structure to our studio. But it also fosters the wish of the students to differentiate from each other, which also creates a wider range of design ideas.

DAVID FRANCO: Yes. The work in the studio is not just deep. It is broad. One way we facilitate this is by looking at many sites each semester – sites with very different climatic and socioeconomic conditions that naturally trigger design differences. Each semester results in different projects, different kinds of approaches to design and technologies. Some connect almost to the vernacular, some incredibly forward-thinking, some almost sci-fi. I think that the reason for that diversity is that we have found a way to empower the students to take their stance on what environmentally responsive architecture is.

GEORGE SCHAFER: The AIA COTE Top Ten for Students competition also helps with this diversity of approaches. The competition describes the 10 criteria in very broad terms so students can easily identify the aspects or components of sustainability or resilience that they're most passionate about and align them with their chosen site. Conversely, the unique characteristics of each program, site and region reveal unique strategies to the students. It is a dialogue between their passions and their discovery. A strength of the way this course leverages the competition is that even though each project must address all of the measures, students have the freedom to prioritize the criteria are most important to their project. This flexibility results in an extraordinarily broad collection of approaches.

One of the things I think that we should probably talk about is how the AIA Framework for Design Excellence measures broaden the conversation about sustainability beyond high-performance buildings to broader themes, such as equitable communities and well-being. A more holistic approach to environmentally responsive architecture, one that this studio has embraced.

DAVID FRANCO: We were three years in when George joined us, and at this time the studio went into the broader field of environmental justice. And it's interesting that it made sense when we added an extra layer of complexity. I have always been extremely interested in politics and architecture, but I think those first two studios the three of us taught together were extremely revealing. Once we brought socioeconomics, new subjects, underserved communities from all kinds of contexts, suddenly, there was this blossoming of new architecture in the studio. But also the themes we chose each year. They created new possibilities to think about the environment and architecture in an incredibly ambitious way that we hadn't seen before. I'm thinking about the projects focused on vulnerable cities and vulnerable populations, which was one of my favorite years.

ULRIKE HEINE: I think there is another aspect that we should add to this whole idea: Communication and collaboration within our studio. Our students are not working as lonely fighters, they work in teams of two or three and we work with them as a collaborating, co-teaching team of professors. We have different backgrounds, different expertise, ensuring that the depth that we're expecting from their projects is fed by all of us, including the students. It is a conversation of equals. I think this is highly unusual in the setting of a studio, which is normally driven by respect and hierarchy. And, of course, sometimes that means we are not on the same page – the professors with each other or with the students.

DAVID FRANCO: It goes back to this idea that we have allowed space. We have created a pedagogical space in which each one's contributions add to the other ones and create a wider discussion. The communication with the students is so open, we give them a voice so they could create ideas that we hadn't suggested or referenced in any way. That is a very healthy dynamic. It is like a client. The client with this exterior entity, the competition. And we are on their side, working a little bit more like consultants, which gives us, ironically, a bit of a secondary role, because the project is really theirs.

And it goes back to what I was saying about our biographies and different interests being impactful. However, many times when I explain, when we get asked about the studio, I answer: "We just do good design." We don't really emphasize environmental technologies. It feels like it's very laissez-faire, like we just let things happen. And it's not true. There's this incredibly complex logistics. We have fifty students in one studio, resulting in twenty-five teams or so, and every desk critique is always with all three faculty together. There's this incredible effort, hours and hours of conversation. It's like we are working in a very implicit way, in a very Socratic way, a pedagogy of massaging and conversation and back and forth, with somewhat clear objectives, but with a huge effort to make that possible.

GEORGE SCHAFER: In a way, there's an interesting tension between the joy of being in such a hyper-collaborative environment, observing students collaborating with each other, constantly collaborating with the faculty that are teaching the course. There's also a necessary discomfort, when we're asking the students to design features of environmentally responsive architecture, starting with thinking about populations, sites, environments and cultures. Conditions that most students in a rural southern state school have not experienced first-hand. We also ask them to work with a partner that is likely a stranger, and also ask them to metabolize feedback from three professors who all bring their own unique perspectives. The students are operating in a very uncomfortable context to develop this really extraordinary work. And this discomfort is really important in a way, because effective collaboration with their partner and metabolizing feedback at a high level of competency requires them to approach design through empathy. This studio throws the students in the deep end of the empathy pool in my view, by acknowledging that resilient design is born out of understanding. And that can be challenging. I think it's also interesting that so many students come out of that experience with such an affection for the semester and can simultaneously articulate how difficult it was at the same time.

ULRIKE HEINE: Well, it's probably for most of the students the first time that they have to please more than one professor. That's a challenge in the first case, especially if those three or four are not necessarily sitting there and agree with each other. But I totally agree that in terms of this discomfort – theme-wise, topic-wise, site-wise – it fosters creativity. We know that there is probably nothing they can draw from because they've either never experienced it before or they've never even thought about it. I think that's something that makes them very challenged designers at that point.

GEORGE SCHAFER: This course was positioned to provide the students with skills for the two semesters where they potentially could go off campus. There is something really beautiful about it particularly in the context of environmental justice, where we are getting these students to think outside of their life experience. It helps prepare many students who have never been out of the state or the country to have that experience abroad. It's setting a foundation for their success when they're experiencing a different culture.

DAVID FRANCO: I'd like to go back to this notion of empathy. We in the profession have been globally aware that after the excesses of the 90s and the 2000s, after the failure of the "starchitect" period, we needed to find new ways for architecture to re-engage with the problems of society. And there have been different approaches to that. A return to evidence-based design, parametrics, a new focus on technical issues or even the social sciences. But I think something that we have experienced in the studio connects with what Peggy Deamer, the architect historian, says when she argues that to change architecture's impact on society, we need to change the practice itself, how architects relate to each other. We need to democratize how offices work. I think this studio is an example of that democratization and illustrates how empathy is critical to the way that faculty collaborate, the way that students are forced to do things and negotiate decisions with other students and with us. It illustrates that from empathy can emerge a totally innovative architecture that has more layers and more connections with society. I think it's a really good example of how the empathy on the smaller scale, on the one-to-one scale affects how we can rethink the real environment as a whole.

GEORGE SCHAFER: The themes confronted by the studio – environmental justice, vulnerable populations, vulnerable environments – these are critical to setting up this culture of empathy. The way we leverage prompts for the competition, the care we take with providing a variety of sites with unique cultural and environmental conditions, are critical to the studio's success.

ULRIKE HEINE: Each semester, even from the beginning, there has been a focus on crises in a certain way. And it's applicable more or less to any place or transferable from one place to another. Sometimes those are crises that are not in the United States, such as the year the project was situated in Spain to address the refugee crisis. All of the studio prompts seem to touch a moment of crisis for humanity – and then we combine these with climatic crises, climatic challenges, where we pick extreme sites. It might be a huge challenge to live in this place or it could be in the future. How can you cope with it and how do you want to design in an environmentally responsive way? It has always been about finding moments of irritation.

DAVID FRANCO: I think there's something important about that human element, whether it's a crisis or an environmental justice topic that's been going on for a long period of time. I'm thinking of the "Lost Spaces" studio that was a conversation about the U.S. interstate highway system and how it disrupted low-income communities or communities of color. A persistent urban design problem that is situated within a social injustice that continues today in America. I think there's something really beautiful about how we embed a project brief into a very human topic.

GEORGE SCHAFER: It is a multi-scale approach to problems. We always tackle deeply local, very personal problems that are implicitly amplified by other problems at global scale. I'm thinking of the "Coastal Tourism as Disaster" studio, and particularly of the St. Helena Island site in South Carolina. There is a historic and ongoing displacement of the Gullah-Geechee community, one that is also exacerbated by the erosion of the barrier island due to sea level rise and extreme weather events. Global forces have local consequences.

DAVID FRANCO: We also learned over time that within a difficult physical context, one that had an exceptional quality or a weirdness to it, better projects would occur. I think that's why the "Lost Spaces" semester, where students were working on sites made problematic by their proximity to elevated highways, design responses emerged from the local conditions and extended to the general condition. In other semesters, design originated from the larger condition, from the climatic crisis and was then infused by the local situation. But there's always this kind of connection between very different scales. And the project has to deal with both. I'm thinking of one of the winning entries from this last year, where their project on the Intracoastal Waterway connected with a very specific infrastructural condition, one that was disjointed and created this programmatic and urbanistic dysfunction. And then the project suddenly connects it to the marshes, with the presence of those marshes in danger, with all of those layers triggered by all of those constraints. The problems trigger something magical, architecturally speaking.

GEORGE SCHAFER: Were multiple sites always leveraged from the very beginning in the studio to help facilitate a broad range of projects?

ULRIKE HEINE: No, in the beginning we had just a single site. And then at some point, well, the classes grew, we had more graduate students than the years before. And it's more entertaining to teach this project on multiple sites. It gives the students the opportunity to very specifically react. If we don't have twenty-five teams working on the same site, they can be very individual in their answers.

DAVID FRANCO: For us, this is also the only rigorous way to discuss climate change. Currently the profession is only talking about floods or hurricanes. We need to talk also about droughts, we need to talk about very different landscapes, very different contexts, heat islands in urban environments. All of these situations are part of the same phenomenon, and it is challenging, but I think by proposing different sites, suddenly you are including that in the collective conversation of the studio.

GEORGE SCHAFER: I'm reflecting on how the exploration of multiple sites and the overall diversity of ideas and approaches also connects to environmental justice being the overarching theme of the studio. For the "Lost Spaces" project, the prompt explicitly framed the environmental justice themes that broadly impacted all of the sites. By contrast, the social, economic and environmental justice issues explored during the "Vulnerable Cities, Vulnerable Populations" and the "Coastal Tourism as Disaster" semesters were identified by the students and came directly from their analysis of the sites. I think there's something beautiful about the fact that the overarching theme of environmental justice is so persistent in this course now, that we don't have to tell students how to approach thinking about their different sites anymore. This really is a testament to how ingrained that overarching concept is to the course that we've developed.

DAVID FRANCO: It certainly has encouraged a culture in the school. While this was happening, the design justice course was created. And different things are happening in our school in Clemson that have made it a really different school when dealing with social justice. I got here eight and a half years ago, and those were not the kind of conversations that we had here, not on those terms. And like you say now, that is completely implicit. We don't need to say it. It is embedded in every site. Students naturally start discussing who are the populations here, who is taking advantage, who do we need, how can we rebalance this and who's being most affected by climate change. This question "Who is the project for?" has become implicitly one of the dominant ones in the studio.

INDEX

Pages in *italics* refer to figures, pages in **bold** refer to tables, and pages followed by "n" refer to notes.